UFOs

UFOs

Generals, Pilots, and Government
Officials Go on the Record

Leslie Kean

Harmony Books / New York

Published in the United States by Harmony Books, an imprint of
the Crown Publishing Group, a division of Random House, Inc., New York.
www.crownpublishing.com

Harmony Books is a registered trademark and the Harmony Books colophon
is a trademark of Random House, Inc.

Library of Congress Cataloging-in-Publication Data

Kean, Leslie.
UFOs : generals, pilots, and government officials go on the record / Leslie Kean. — 1st ed.
p.   cm.
1. Unidentified flying objects—Research. 2. Unidentified flying objects—Sightings
and encounters. 3. Public records. I. Title.
TL789.K357 2010
001.942—dc22
2010015221

ISBN 978-0-307-71684-2

Printed in the United States of America

Design by Leonard W. Henderson

10 9 8 7 6 5 4 3 2

For Paul

CONTENTS

FOREWORD
By John Podesta

As someone interested in the question of UFOs, I think I have always understood the difference between fact and fiction. I guess you could call me a curious skeptic. But I'm skeptical about many things, including the notion that government always knows best, and that the people can't be trusted with the truth. That's why I've dedicated three decades of my life, in private practice, as counsel to the Senate Judiciary Committee, at the White House under President Clinton, and now with the Center for American Progress, to the fundamental principle of protecting openness in government.

Because of this commitment, I have supported the work of investigative journalist Leslie Kean and her organization, the Coalition for Freedom of Information, in their initiative, launched in 2001, to obtain documents about UFOs through the Freedom of Information Act. In the spirit of inquiry, Kean successfully sought an injunction in federal court on one important case, as was her right under the law.

The time to pull the curtain back on this subject is long overdue. *UFOs: Generals, Pilots, and Government Officials Go on the Record* involves just such an effort, and it appeals to open-minded people such as myself. Presenting the facts, the book includes statements from only the most credible sources—those in a position to know—about a fascinating phenomenon, the nature of which is yet to be determined. Kean and her impressive team of contributors make no untoward claims, but provide a rational analysis of the most pertinent information, much of it presented here firsthand in riveting detail, stating that further investigations are needed. Kean has more than done her homework as a dogged investigative reporter, diligently contending with this perplexing subject for ten years while having to face attitudes of ridicule and denial within government. Yet she persevered, and her book clearly leaves the taboo against taking UFOs seriously with no leg to stand on.

Kean and her distinguished co-writers call for the establishment of a small U.S. government agency to cooperate with other countries that are already formally investigating, reviewing, and releasing information relevant to UFOs. This new agency would handle release of documents and any future investigations with openness and efficiency. It's an idea worth considering, and it is definitely time for government, scientists, and aviation experts to work together in unraveling the questions about UFOs that have so far remained in the dark. It's time to find out what the truth really is that's out there. The American people—and people around the world—want to know, and they *can* handle the truth. *UFOs: Generals, Pilots, and Government Officials Go on the Record* represents a pivotal step in that direction, laying the groundwork for a new way forward.

UFOs

INTRODUCTION

Ten years ago, as an investigative reporter working for a California public radio station, I was suddenly confronted with a seemingly impossible reality. A colleague in Paris sent me an extraordinary new study by former high-ranking French officials documenting the existence of unidentified flying objects and exploring their potential impact on national security. Now known as the COMETA Report, this unprecedented white paper marked the first time in any country that a group of this size and stature had declared that UFOs—solid but as yet unexplained objects in the sky—constitute a real phenomenon warranting immediate international attention.

The distinguished COMETA authors—thirteen retired generals, scientists, and space experts working independently of the French government—had spent three years analyzing military and pilot encounters with UFOs. In the cases they present, all conventional explanations of something natural or man-made had been eliminated by the authors and their associated teams of experts, and yet these objects were observed at close range by pilots, tracked on radar, and officially photographed. They achieved tremendous speeds and accelerations, made sharp, right-angle turns in a flash, and could stop and stand still in midair, seeming to defy the laws of physics. What could this mean? Since some of the military officers on the COMETA panel were serving with the French Institute of Higher Studies for National Defense, a government-financed strategic planning agency, their characterization of UFOs as a phenomenon with possible national security implications assumed a grave importance.

In their ninety-page report, written with objectivity, clarity, and logic, the authors explained that about 5 percent of sightings—those for which there is enough solid documentation to eliminate other possibilities—cannot be easily attributed to earthly sources, such as secret military exercises or natural phenomena. This 5 percent seem to be "completely unknown flying machines with exceptional performances that are guided by a natural or artificial intelligence." In its startling conclusion, the authors state that "numerous manifestations observed by reliable witnesses could be the work of craft of extraterrestrial origin." In fact, they wrote, the most logical explanation for these sightings is "the extraterrestrial hypothesis."

This did not mean that they accepted this conclusion as fact or had any particular beliefs about it one way or the other. They made very clear that the nature and origin of the objects remain unknown. By "hypothesis," the authors simply meant an unproved theory, a *possible*, plausible explanation that needed to be tested before it could be decided, but remained only a thesis until that happened. However, the conviction with which they put forth this theory as the "most likely" solution to the puzzle, since others had been ruled out in so many cases, was provocative. Official data about UFOs from around the world was accessible to the members of the group, and they were determined to respond rationally, avoiding prejudice. They did so without reserve.

Who were the people making these statements? Among them, all retired, were a four-star general, a three-star admiral, a Major General, and the former head of the French equivalent of NASA. It was their credentials that made the report worthy of serious consideration. Other military officers, engineers, scientists, a national chief of police, and the head of a government agency studying the phenomenon completed the impressive contributing group. The study was not a government-sanctioned one, but was undertaken independently, and then presented to the highest levels of government in France.

The foreword states that the report "contributes toward stripping the phenomenon of UFOs of its irrational layer," and indeed, the study achieved its goal. Yet the group arrived at a determination that most government officials and scientists in the United States would still consider far-fetched. Meanwhile, everyone agrees that if these UFOs *were* proven to be probes or vehicles from outside Earth, that would be a monumental development in human history, a milestone in the evolution of civilization. If there was even a slight possibility of such a discovery, I thought, it seemed well worth the effort for scientists to try to find out. And here was a highly respectable group from a sophisticated European country stating that such an outcome was a plausible and even likely expectation.

This explains why and how I first became interested in the issue of UFOs, the question of what we actually do and don't know about them, and how we might find out more. The COMETA Report was a catalyst. As much as I may have wanted to, it was hard for me to let it go, to simply return to my regular work and set it aside. I kept wondering, could there really be technological objects flying around that are not man-made? Couldn't these craft possibly be highly secret American constructions, or advanced military test craft from some other country? No, said the generals and the rest of the high-level French panel. Countries do not fly experimental aircraft repeatedly in foreign airspace without informing the host country and then lie about it later. As I dug deeper, I learned that these objects have appeared for decades in a variety of shapes and sizes, sometimes in flaps or "waves," all over the globe, demonstrating capabilities beyond our scientific understanding. This was not a myth. And maybe, I thought, the French generals and their colleagues knew even more than they disclosed.

Not only did all the members stand by the conclusion, they also urged international action. The writers recommended that France establish "sectorial cooperation agreements with interested European and foreign countries" on the matter of UFOs, and that the European Union undertake diplomatic action with the United States, "exerting useful pressure to clarify this crucial issue which must fall within the scope of political and strategic alliances." The report, titled "UFOs and Defense: What Should We Prepare For?," is most fundamentally a call to action, a request for preparedness in anticipation of future encounters with the unknown objects.

I had no idea where all this might lead—for me, for any government, or for our future.

My French colleague called to follow up and explained that he had surreptitiously slipped me an advance English copy of the report, just translated. The news was being held for a later release, and so far the report had been published only in France. My friend knew that I was an open-minded freelance reporter with ties to many publishing outlets, and he wanted me to get a head start on the story rather than leave it to the conventional mainstream media, which rarely took UFOs seriously. "You are the only reporter in all of America to have the English version," he told me excitedly over the phone from Paris. "It's all yours. But don't let anyone know where you got it."

The challenge was both enticing and nerve-racking. Secretly, I started to look into the UFO subject more extensively, without telling any of my otherwise close colleagues at the radio station. I knew that I was exploring something most journalists considered ridiculous, or titillating at best, but otherwise irrelevant to the life-and-death struggles of human beings, issues that should be the focus of any responsible, progressive reporter. As the months passed and I became increasingly concerned about keeping my expanding interest quiet while producing and hosting a daily investigative news show, I began to feel as if I were covering up something shameful and forbidden, like the use of an illegal drug. In retrospect, the intensity of my worry and insecurity was overblown, but the taboo regarding UFOs had power over me, and it took a while before I felt armed with enough facts and insight to handle the attitudes of those I worked with so compatibly in every other respect.

This was not an easy subject to take on, and I understand why other journalists haven't done so. At first, I felt burdened by what seemed to be almost insurmountable obstacles. The UFO story was journalistically elusive, contaminated by conspiracy theories, disinformation, and just plain sloppiness, all of which had to be carefully separated from the legitimate material. The questions raised by the UFO phenomenon were deeply disturbing to our accustomed ways of thinking. The subject carried a terrible stigma and was therefore a professional risk for those publicly engaged with it. But it also pointed to something possibly revolutionary, something that could challenge our entire worldview. Although frightening, that made it all the more appealing to me, I have to confess. And the more I learned, the better I understood the validity of additional case studies and government documents shedding light on the UFO question. The aggregate data, the accumulation of evidence over decades, was utterly compelling and completely mystifying. Despite the problems, there was simply no way I could force myself to ignore it.

As it turned out, that unsolicited report from France radically changed the course of my career as a journalist in ways I never could have imagined at the time. UFOs became the focus of my professional life after the publication of my first story about them in the *Boston Globe*. The editor of the *Globe*'s Sunday Forum, a weekly news analysis section in which I had published previously, was apprehensive about covering the subject of UFOs. It understandably unnerved her, but after much discussion she was courageous enough to run my lengthy story. I was extremely nervous about "coming out of the closet" professionally as a reporter who—God forbid!—found this silly subject worthy of attention. But I knew this was a scoop, and how could I resist? I broke the news of the COMETA Report, just as my French colleague had requested six months before, and the stature of the generals and others authoring the report carried the day, exempting me from ridicule. I even included additional analysis based on revelatory information spelled out in official U.S. government documents pertaining to UFOs and national security, all of which backed up the French perspective. To my delight, the article was distributed through the *New York Times* wire service and picked up by newspapers across the country. Clearly, there was national interest.

People following the UFO subject were elated that at least one prestigious newspaper had taken the story seriously, and a congressional staffer even sent a complimentary letter to the *Globe*. I received numerous e-mails from witnesses to UFO events in response to the article, including a few pilots, who had so far never dared to come forward. My eyes were opened by this, and I had now crossed the point of no return.

The story carried that disquieting quote—printed in black and white, clear as the other stories of the day with which it oddly mingled—about "completely unknown flying machines with exceptional performances that are guided by a natural or artificial intelligence," as described by the retired French officials. I naively thought this would *have* to generate some kind of news buzz, and that other journalists would eagerly jump in to pick up where I had left off. I knew there was disdain for UFOs in the culture, but I also knew that this was a breaking story passing muster with a leading mainstream paper. Amazingly, *nothing happened.* I had been exposed to another aspect of this strange world. It was the beginning of a rude awakening, a rite of passage into the perplexing reality that UFOs cannot be acknowledged at all, even as simply the unidentified flying objects that they are. It was as if everyone was pretending that they didn't exist.

Ever since the *Globe* story solidified my interest and increased my confidence, I have been focused on investigating and coming to terms with this subject—a process that never ends. Fundamentally, after many years of research and in-depth interviews with key players, I have learned that UFOs are a genuine scientific mystery. There have been extraordinary UFO sightings occurring in America for more than sixty years, many by pilots and military personnel, and many yielding physical evidence. Volumes of case studies have been published by qualified researchers and scientists since the 1950s, documenting UFO incidents all over the globe and leaving a solid record begging for further analysis by contemporary scientists.

The most credible sources clearly recognized, and stated repeatedly, that we don't yet know what the objects are—contrary to public assumptions that UFOs, by definition, are extraterrestrial spaceships. But I had to come to terms, over and over again, with the fact that these amazing, high-performance unidentified objects *do* exist, without question—just as the COMETA authors had unequivocally stated. There is enough data available to make this clear to anyone who decides to take the time to study it. Because that alone was so potentially explosive, I couldn't quite understand the indifference it generated among those who took it seriously enough to rise above ridicule but who nonetheless remained blasé and disinterested.

Eventually I came to realize—repeatedly, through the research and publication of my subsequent stories, each of which seemed like earth-shattering news to me then but was never enough to stimulate change—that the UFO story could *not* be properly told, nor could the taboo be overcome, through any one short news piece, no matter how many there were. I now believe that the only way to adequately convey the full story—to really break the news about the existence of UFOs and convey the impact of the material for the person so far unexposed to it— is through a book such as this one, which includes some of the world's best sources speaking in detail for themselves. Sound bites and short quotes cannot carry a story of this magnitude.

The chapters you are about to read will address the fundamental questions about UFOs that concern so many people. What do we *really know* about them? Is it actually possible that some of these objects are from outer space? Do pilots ever see them? How do governments and militaries handle sightings? Why is there so much ridicule and denial about UFOs in America? The answers, on all counts, are nothing short of astonishing.

As any journalist would, I have relied on official sources, documents released through the Freedom of Information Act, corroborated case reports, physical evidence, and numerous interviews with military and aviation witnesses and government investigators from around the world. I've come to know many of these official witnesses personally, and have no doubt as to the credibility of their accounts, which are almost always corroborated by others. Some have conveyed information, and showed me documents, that must remain off the record because of their sensitivity, and other such documents, provided by very trustworthy sources, cannot be verified or corroborated but are still valuable as background. I have also met, interviewed, and come to know numerous civilian witnesses over the years, regular people from all walks of life, who have impressed me with their sincere and clear accounts of amazing UFO incidents. They, too, have made essential contributions to the search for understanding about the phenomenon.

My role here is to write as an objective observer, and as a guide. At the same time, I take a position in support of an effort to solve the many unresolved questions about UFOs, rather than ignoring them, and in support of the witnesses and experts who have come forward. In so doing, I'm directly and openly confronting irrational attitudes and misinformation. This means I'm practicing a form of "advocacy journalism," something that I've never objected to and that is the modus operandi for many investigative reporters who dig into a story to serve a greater cause. I'm not a "believer" in anything except respect for the facts, even when they don't conform to our established worldview. The UFO issue is so unorthodox that even a straightforward, rational approach can seem as though it's crossed a line into questionable territory. I've done my best to keep all this information clear, logical, and well-documented.

That is why much of this book consists of personal accounts from expert investigators and witnesses who will address the UFO issue directly, some for the first time. Through their words, readers will be given firsthand access to the material, and can arrive at their own informed conclusions.

These individuals from nine countries are highly trained men who were assigned the daunting job of confronting this phenomenon through intimate investigation, or who directly witnessed it, not by any choice of their own. Some of them have been given access to secret files, insider witnesses, and unfolding case investigations way beyond the reach of any journalist or anyone outside of their closed, privileged world. They are coming forward collectively here, to give all of us access, and to explain what they know about UFOs, in their professional capacity as pilots, government officials, and high-ranking military officers.

On a personal level, each one has been transformed in one way or another, sometimes drastically so, by this interaction with the "impossible." They are all baffled and want answers to the same serious questions as the rest of us, but usually for their own reasons. Each one began his relationship to the UFO issue as a natural skeptic, and even though many are now retired from their official jobs investigating UFOs, most have not been able to disengage from the intense drive to want to find out what UFOs are. They remain involved in various ways. One is planning to teach a course on UFO history at a prominent university; another is contacted frequently by the media to be a spokesperson on the issue; a former NASA scientist heads a research group studying anomalous aerial phenomena; a former government investigator is often called via cell phone by nervous Air Force personnel observing strange phenomena in remote locations. So in this sense, these men are not really fully "retired." And some are now captains working for commercial airlines.

I noticed that many, even the ones I came to know well, were hesitant to reveal the emotional aspect of their experiences dealing with UFOs. Some witnesses struggle for years with the impact of a mind-boggling close encounter. It was my job to nudge as much as I could from the minds of these reticent military men and Air Force pilots not prone to disclosing their fears. These are men oriented toward duty first, and the significance of their statements cannot be overemphasized. This courageous group is breaking a huge story for the world.

Over many years they have all discovered a great deal about UFOs, despite the phenomenon's ability to remain unidentified even while making repeated, tantalizing appearances in so-called waves, or engaging in cat-and-mouse chases with Air Force pilots. The objects come and go, sometimes leaving a blip on radar, an image on film, or an imprint on the ground. This diverse group can provide as intimate and factual a look at this mysterious phenomenon as we can ever hope to acquire as outsiders.

None of these writers were privy to the others' statements, nor, to my surprise, have any ever asked me what the other contributors were writing about. Even so, striking similarities exist, not only in their reports of the UFOs themselves, but also in their interpretations, attitudes, and ideas for future resolution. To me, this uniformity validates the worldwide nature of the phenomenon, and it also shows that when properly investigated, the same conclusions are drawn no matter where the investigation takes place.

There exists a universal curiosity, increasing over time, about the UFO mystery. I have seen it grow, and have observed an improvement in straightforward media coverage about UFOs since I began this study ten years ago. The more we learn, the more confounding it becomes. Still, many people continue to think the subject is based on fantasy or mistaken identity, or is some kind of a joke and therefore a waste of time. My deepest hope is that these people in particular will read through this entire book, sticking with it from beginning to end, and then draw a conclusion. We can all agree, I assume, that no one is entitled to dismiss a subject without knowing something about it.

I have done my best to distill from a huge mass of material some of the most compelling and essential facts. In this country, UFOs became a national issue in the late 1940s, when there were many sightings of great public interest and concern that were covered widely by the media. The U.S. Air Force took the lead in addressing these events, complicated by the onset of the Cold War, attempting to explain away as many cases as possible in order to divert public attention from the mystery. Behind the scenes, the topic was of great concern at the highest levels, and the Air Force was not equipped to protect the public from an entirely unknown but apparently technological phenomenon that could come and go at will. In the early 1950s, it established Project Blue Book, a small agency that received reports from citizens, investigated the reports, and offered explanations to the media and the public. Blue Book gradually solidified as largely a public relations effort intent on debunking UFO sightings. Hundreds of files accumulated, and the Air Force closed down the program in 1970, ending all official investigations—or so they said publicly—without having found an explanation for many shocking UFO incidents. The cases presented by our contributors all occurred after the close of Project Blue Book, between 1976 and 2007.

Our government still stays out of the UFO controversy and has no policy in place to address growing concern. Within the historical framework, the upcoming chapters will examine the role of the CIA in establishing the protocol for the debunking of UFOs; the stark contrast between the handling of UFOs by our own government and the governments of other countries; the issues of aviation safety and national security as they pertain to UFO incidents; the psychology of the UFO taboo; and the question of a U.S. government cover-up.

Much of the American public has grown increasingly frustrated with the pattern of government denials about UFOs, especially as the evidence has mounted over time. With digital cameras and cell phones now commonplace, UFO photos are snapped almost every day, although they are easier to fake, making the new technology a mixed blessing. As exoplanets are discovered and scientists acknowledge the probability of life elsewhere in the universe, the demand for studying the neglected UFO phenomenon has become imperative. I think you will agree, by the time you finish reading, that there is now renewed hope for solving the UFO enigma, and that you will also agree as to the signal importance of that endeavor.

Defining the Indefinable: What Is a UFO?

It's extremely important to establish at the very beginning that neither I nor the other writers are claiming that there are alien spacecraft in our skies, simply because we do not deny data showing a physical presence of *something* there. The term "UFO" has been misused and has become so much a part of popular culture that its original (and accurate) definition has been nearly completely lost. Almost everyone equates the term "UFO" with extraterrestrial spacecraft, and thus, in a perverse twist of meaning, the acronym has been transformed to mean something *identified* rather than something *unidentified*. The false but widespread assumption that a UFO is, of necessity, an alien spaceship is usually the reason the term generates such an exaggerated and confusing range of emotional responses. A recognition of the extraterrestrial *hypothesis* as being a valid, although unproved, possible explanation worthy of further scientific scrutiny is something entirely different from approaching the subject of UFOs as if this discovery had already been made.

Historically, it was the U.S. Air Force that, some fifty years ago, invented the term "unidentified flying object" to replace the popular but more lurid phrase "flying saucer." The Air Force defined a "UFO" as "any airborne object which by performance, aerodynamic characteristics, or unusual features does not conform to any presently known aircraft or missile type, or which cannot be positively identified as a familiar object." This is the definition embraced by all the contributors to this book, and the definition employed by all relevant government documents and official pilot reports.

If an object in the sky cannot be identified but we still can't rule out the possibility that it *could* be if we had more data, then it is *not* a true unknown. In that situation, we can't determine either what it is or what it is not. Again, a genuine UFO, the UFO we are concerned with in this book, is an object that, for example, exhibits extraordinary capabilities beyond known technology while being documented on radar and observed by multiple qualified people, to such an extent that enough data is obtained and enough study is undertaken to eliminate other known possibilities.

Because there is so much baggage associated with the term "UFO," some scientists and other experts have employed a new terminology to separate serious studies from the more frivolous. Instead of "UFO," some of our contributors have chosen to use "unidentified aerial phenomena" or "UAP," which can be used in both the singular (for the phenomenon) and the plural. Richard Haines, former NASA senior scientist and aviation safety expert, defines UAP as:

> The visual stimulus that produces a sighting report of an object or light seen in the sky, the appearance and/or flight dynamics of which do not suggest a logical, conventional flying object and which remains unidentified after close scrutiny of all available evidence by persons who are technically capable of making both a full technical identification as well as a common-sense identification, if one is possible.

In the context of this book, the terms UFO and UAP mean essentially the same thing and will be used interchangeably, although some writers prefer to use one or the other exclusively. "UAP" suggests a broader scope, incorporating perhaps a wider range of phenomena, which, for example, may not appear to be a flying object. No matter which acronym is employed, the phenomenon is often motionless or hovering, not flying, and sometimes is simply seen as unusual lights rather than a solid object, especially at night when brilliant illumination overpowers the observation of any physical structure. "UAP" maintains the clarity that these unusual objects and lights may represent many types of phenomena originating from different sources.

A second fundamentally important point is that roughly 90 to 95 percent of UFO sightings *can* be explained. Within the remaining 5 to 10 percent, once an object has been determined to be a genuine UFO by the proper standards, then all we know is what it is *not*: something man-made or natural, or an outright hoax, of which there are unfortunately too many. Examples of phenomena sometimes mistaken for UFOs are weather balloons, flares, sky lanterns, planes flying in formation, secret military aircraft, birds reflecting the sun, planes reflecting the sun, blimps, helicopters, the planet Venus or Mars, meteors or meteorites, space junk, satellites, sundogs, ball lightning, ice crystals, reflected light off clouds, lights on the ground or lights reflected on a cockpit window, temperature inversions, hole-punch clouds, and the list goes on! Yes, the vast majority of reports can usually be explained by one of the above, but of course it's only the ones that can't that we're interested in.

It follows, therefore, that the often asked question "Do you believe in UFOs?" is actually baseless, but it's frequently asked and creates endless problems in communication. It really doesn't make sense, because we know that *unidentified* objects exist, officially documented and defined as such by the U.S. Air Force and other government bodies around the world. For over fifty years, the reality of unidentified flying objects has not been a question of belief or a matter of faith, opinion, or choice. Rather, when using the correct definition of UFO, it is a matter of *fact*. Like conventional *identified* objects—such as aircraft, missiles, and other types of man-made equipment—these *unidentifieds* can also be photographed, create radar returns, leave marks on the ground, and be observed and described by multiple independent witnesses at separate locations. In terms of belief, the questioner is *really* asking, "Do you believe in alien spaceships?" That is an entirely different question.

To approach UFOs rationally, we must maintain the agnostic position regarding their nature or origin, because we simply don't know the answers yet. By being agnostics, we are taking a giant step forward. So often, the UFO debate fuels two polarities, both representing untenable positions. On one side, the "believers" proclaim that extraterrestrials have arrived from outer space and that we already *know* that UFOs are alien vehicles, and on the other, the "debunkers" argue with aggressive defensiveness that UFOs don't exist at all. This counterproductive battle has unfortunately dominated public discourse for a long time, only heightening confusion and creating more distance from the scientific—the agnostic—approach.

Principled skepticism is the foundational premise of this book. Astrophysicist Bernard Haisch, former science editor for *The Astrophysical Journal* and *The Journal of Scientific Exploration,* defines a true skeptic as "one who practices the method of suspended judgment, engages in rational and dispassionate reasoning as exemplified by the scientific method, shows willingness to consider alternative explanations without prejudice based on prior beliefs, and who seeks out evidence and carefully scrutinizes its validity." I invite you to look at the material presented here from the perspective of an agnostic—objectively, with an open and truly skeptical mind.

Now we can begin a fascinating journey. I will present some of the most powerful material that so profoundly impacted me during my own process of exploration and discovery. During that process, the other writers and I ask the reader to consider the veracity of the following points, to be revisited at the end of book, which I have distilled from my ten years of looking into the UFO subject. These five premises are thoroughly evaluated and illustrated by the evidence throughout the volume:

(1) There exists in our skies, worldwide, a solid, physical phenomenon that appears to be under intelligent control and is capable of speeds, maneuverability, and luminosity beyond current known technology.

(2) UFO incursions, often in restricted airspace, can cause aviation safety hazards and raise national security concerns, even though the objects have not demonstrated overtly hostile acts.

(3) The U.S. government routinely ignores UFOs and, when pressed, issues false explanations. Its indifference and/or dismissals are irresponsible, disrespectful to credible, often expert witnesses, and potentially dangerous.

(4) The hypothesis that UFOs are of extraterrestrial or interdimensional origin is a rational one and must be taken into account, given the data we have. However, the actual origin and nature

of UFOs have not yet been determined by scientists, and remain unknown.

(5) Given its potential implications, the evidence calls for systematic scientific investigation involving U.S. government support and international cooperation.

I believe that after reading this book, the discerning reader will accept—or at least acknowledge as plausible—these five positions, as remarkable or even inconceivable as they seemed at the outset.

Leslie Kean
New York City

PART 1

OBJECTS OF UNKNOWN ORIGIN

"All truth passes through three stages. First, it is ridiculed, second, it is violently opposed, and third, it is accepted as self-evident."

ARTHUR SCHOPENHAUER

CHAPTER 1

Majestic Craft with Powerful Beaming Spotlights

W e begin this exploration on very solid ground, with a Major General's firsthand chronicle of one of the most vivid and well-documented UFO cases ever. What you are about to read will demonstrate the dramatic, and very mysterious, physicality of UFOs—in this case, ones that were unusually bold. Although parts may sound like science fiction, they are not. The fact is that silent gliding or hovering objects, usually triangular, were seen by thousands of people and investigated by university scientists and government officials, yet they could never be explained. They left imprints on film, and although virtually impossible to detect on radar, they triggered the launching of Air Force F-16s in anxious pursuit. The sightings occurred in a more than two-year "wave" over Belgium, beginning in late 1989.

To launch this book's exploration into the UFO phenomenon, Belgian Major General Wilfried De Brouwer, now retired, has provided an exclusive account that includes some personal responses he has never expressed before. As chief of the Operations Division in the Air Staff, then Colonel De Brouwer played a prominent role, along with officials from other branches of government, in mobilizing various departments to try to identify the strange intruders that kept showing up unannounced over cities and countryside. "Hundreds of people saw a majestic triangular craft with a span of approximately a hundred and twenty feet and powerful beaming spotlights, moving very slowly without making any significant noise but, in several cases, accelerating to very high speeds," De Brouwer stated publicly a few years ago, describing only the first night of the wave. Numerous police officers were among the initial group of witnesses, reporting from different locations as the multiple flying craft hovered and glided and lit up fields along their routes—the same officers who had joked dismissively when first receiving radio calls about the sightings. And the strange objects kept returning, for some unfathomable reason, to display themselves over the otherwise quiet territory of Belgium.

Colonel De Brouwer was tasked to handle the UFO wave by his country's defense minister, Guy Coëme. After spending twenty years as a fighter pilot in the Belgian Air Force, De Brouwer had been appointed to the Strategic Planning Branch in NATO in 1983, while a colonel. He then became Wing Commander of the Belgian Air Force Transport Wing and, in 1989, chief of the Operations Division in the Air Staff. Promoted to Major General in 1991, he became Deputy Chief of Staff of the Belgian Air Force, in charge of operations, planning, and human resources. Beginning in 1995, after his retirement from the Air Force, he worked for more than ten years as a consultant for the United Nations to improve the UN Logistics rapid-response capabilities during emergencies. A man of great integrity and responsibility, De Brouwer was determined to do everything he could to find out what was invading Belgian airspace and repeatedly committing infractions of basic aviation rules.

I first came to know General De Brouwer personally when arranging his trip to Washington, D.C., in November 2007 to speak at an international press conference I organized with filmmaker James Fox. We brought together a panel of former high-ranking government, aviation, and military officials from seven countries to speak to the press about UFO incidents and investigations, which was filmed for a new documentary. We also wanted to give these courageous speakers the opportunity to meet their counterparts from other countries and talk privately over a period of days. Many of the contributors to this book met then for the first time.

General De Brouwer is extremely concerned about factual accuracy, conservative in his estimations, and meticulous in his attention to detail. He is a man who does not jump to conclusions, nor is he prone to exaggeration or embellishment. His concern for safeguarding the accurate record of events in Belgium has not let up, despite the passage of time. "Recently, when on the Internet, I discovered an accumulation of misinformation about the Belgian UFO wave," he wrote me in an e-mail while we worked on editing his extensive text. "This incited me to react; I cannot accept that so-called researchers come forward with assumptions that are based on incorrect information. Testimonies of hundreds of people are neglected and attempts are made to convince outsiders that the observations were nothing more than misperceptions of ordinary craft. Also, the official statements of the Minister of Defense and the Air Force have been neglected or misinterpreted by these 'researchers.' "

In one of our more recent conversations, I asked the general to reflect back on his experience during the Belgian UFO wave twenty years ago—which he says was unique but also frustrating, since they were unable to identify the trespassing craft. What impressed him most was the utmost sincerity of the witnesses he spoke to, many of whom were "highly qualified intellectuals genuinely overwhelmed by what they had seen and convinced that they were not dealing with conventional technology." Unfortunately, they were often afraid to come forward because of the stigma attached to UFOs. "One person I had known for many years worked within NATO at the time," De Brouwer said. "He was so astonished that he didn't dare to mention it to anyone, not even to his wife. He only conveyed his experience to me on condition that I wouldn't reveal his name."

I had the good fortune of conversing with one highly placed, expert witness who did not restrain himself, despite the risks. Colonel André Amond, a retired civil engineer, was the director of military infrastructure for the Belgian Army and also formerly in charge of army environmental-impact issues at the Joint Staff level, cooperating closely with American officials. As De Brouwer reports in the next chapter, Amond and his wife had an extensive look at one of these low-flying machines while driving down a country road and parking along the side. Amond had absolutely no doubt about the exceptional nature of what he saw. With total conviction, he went all the way to the top, filing a written report and providing a series of drawings for the Belgian Defense Minister.

As far as he's concerned, Colonel Amond was able to eliminate all possible explanations for the object, and states that it was some kind of "unknown aerial vehicle." As for his reflection on this event two decades later, he wrote in an e-mail: "Today there is not yet any explanation! That is a pity because I want to know before dying. Give me a correct explanation of my sighting; that is all that I can ask." He speaks for thousands of others who never thought about UFOs before having the unasked-for experience of seeing one. For many, the effect of that lasts a lifetime.

In order to fully grasp the significance of the evidence to be presented by General De Brouwer, we must recognize the special circumstances of this extraordinary series of events. Most UFO cases are not "waves," and don't offer up nearly as much data as this one did. Usually they involve a one-time incident, and naturally these are harder both to document and to investigate. The many hundreds of vivid and consistent case reports collected over time in Belgium—accumulated and investigated by a group of scientists working with the Air Force—created opportunities for radar detection and other technical applications that benefit from advanced preparation. The large number of sightings increased the likelihood of obtaining valid photos and video footage. The military had adequate time to assess and test a range of options for what the objects might be, which could be either verified or eliminated based on official inquiries, such as whether any helicopters were airborne at a particular time. Officials could prepare for future visits of the UFOs by training radar specialists to handle these exceptional targets and readying Air Force jets to launch at a moment's notice. As events unfolded over months and years in Belgium, all mundane, conventional explanations were ruled out. It became very clear what the objects were *not,* but there was still no clarity about what they *were.*

Eventually, the only possible option left, no matter how remote, was that the objects were F-117A stealth fighters or other secret American military aircraft, sent out on some kind of experimental, clandestine exercise. General De Brouwer thought it extremely unlikely that secret aircraft would be sent to fly repeatedly over Belgium without any official notification, in violation of air rules, since no U.S. Air Force overflight requests had been received. He was also aware that the technological abilities the objects displayed were beyond the capacity even of experimental aircraft—which, the general points out, remains the case today. Nonetheless, he made inquiries to the U.S. Embassy in Brussels, and to other NATO partners through informal contacts with their attachés.

The answer was exactly what he expected. And the results of his inquiry are spelled out in a U.S. government document, classified at the time, but since released through the Freedom of Information Act. The March 1990 memo "Belgium and the UFO Issue" notes that De Brouwer asked whether the objects were American B-2 or F-117 military aircraft, stating that he made the inquiry despite his clarity that "the alleged observations did not correspond in any way to the observable characteristics of either U.S. aircraft." The document further states that "the USAF did confirm to the BAF [Belgian Air Force] and Belgian MOD [Ministry of Defense] that no USAF stealth aircraft were operating in the Ardennes area during the periods in question." De Brouwer reported to me that he was also assured privately by an American official that the U.S. had no "black program" that could have caused these multiple sightings.

In 1992, Belgium's defense minister, Leo Delcroix, confirmed this once again when replying to a letter from a French researcher. "Unfortunately, no explanation has been found to date," he wrote. "The nature and origin of the phenomenon remain unknown. One theory can be definitely dismissed, however, since the Belgian Armed Forces have been positively assured by American authorities that there has never been any sort of American aerial test flight."

This is an important point to keep in mind when reading the witness accounts provided by De Brouwer. We're stuck with a serious dilemma. Has the military from some country been testing new, extremely advanced craft since the mid-1970s, which is when reports of such triangular craft began? Was Belgium selected as the site for repeated test flights, monitored from a secret base somewhere? Common sense tells us that if a government had developed huge craft that can hover motionless only a few hundred feet up, and then speed off in the blink of an eye—all without making a sound—such technology would have revolutionized both air travel and modern warfare, and probably physics as well. In the two decades since the Belgian wave, the United States has been involved in three wars; had we possessed such advanced capabilities, they would surely have been put to use by now. If some government *was* secretly, and inexplicably, flying this marvel over Belgium, it would have had to lie to the Belgian authorities when inquiries were made and thus disrupt the partnership among NATO members, which is based on mutual respect and trust. And *every* person involved with the creation and flight of this highly advanced craft would have had to have kept the miraculous technology and its repeated test flights secret—indeed, no one has come forward and nothing about such an enterprise has ever leaked out. Nonetheless, in the minds of some, this will remain as a possibility, no matter how unlikely.

As far as General De Brouwer is concerned, that possibility has been completely ruled out. So, to his mind, what's left? "I am approaching the UAP issue in a pragmatic way. I stick to the facts and avoid extrapolations to possible extraterrestrial activities," General De Brouwer responded by e-mail. "Nevertheless, I encourage scientific research which should be based on the objective analysis of a number of observations reported during the Belgian wave. Such research should not exclude the extraterrestrial option."

Lastly, I want to point out the significance of the close-up color photograph of an unidentified object that De Brouwer will present—one of the most revealing UFO images of all time. Readers might reasonably ask why there aren't more unequivocal pictures and videos of the Belgian objects, since there were so many sightings. Partly, this was because of the strict requirements of the authorities regarding the acceptance of photographs; their screening methods eliminated all questionable and unverifiable images. In addition, it's easy to forget that twenty years ago, cell phones and relatively inexpensive, consumer-level digital and video cameras were not yet in use. Most often, people did not have loaded cameras handy at the unpredictable times when UFOs passed overhead, such as at night while driving. In my conversations with many UFO witnesses over the years, I've learned that when observing something as awesome, and sometimes frightening, as a gigantic low-flying UFO, people become almost transfixed. They are seeing something that isn't supposed to exist, something ominous, huge and silent, that was previously unimaginable. Most do not take their eyes off the otherworldly thing except maybe to quickly summon family members or friends within earshot. They keep staring, and the distraction of taking a picture is not on their minds. The craft is usually moving away, soon to be out of sight. They do not want to run inside the house to look for a camera, or unpack a bag in the trunk of the car to find one, or worry if it's loaded. The moment is too unusual, too breathtaking.

Even when a picture *is* taken, it doesn't always come out. If the lights are some distance away and the exposure too short, nothing shows up in the frame. Also, other characteristics of the UFO can inhibit the registering of its bright lights on film. In one case, a Belgian movie producer and two colleagues, using high-sensitivity film, photographed one of the objects passing directly overhead. The photographer estimated its altitude to be only about 1,000 feet, with the object's diameter six times that of a full moon. As a control, he photographed an ordinary airplane several minutes later in the same spot, using all the same settings on the camera.

On the pictures, however, the bright "spotlights" on the UFO, which to the viewers' eyes had looked much, much brighter than the lights on the airplane, were hardly discernible. The triangular shape of the UFO, clearly visible to the naked eye, was also lost on the film. At the same time, the airplane lights came out brighter than those of the UFO, appearing just the way they had looked from the ground, even though the UFO was much closer to the observers than the airplane. Laboratory experiments show that this was likely due to the effect of infrared light around the UFO, which can cause even such an object to disappear altogether in a photograph. This could be one reason why so few usable pictures were received by investigators during the Belgian wave, and why bona fide UFO pictures are not as common as one might expect.

Witness drawings have an important role to play, encapsulating details imprinted in the memories of observers immediately after their sightings. Investigators can then make comparisons between renditions made in different locations at different times, or by multiple witnesses to the same event from different vantage points—all by people who don't know each other. "The day will come, undoubtedly, when the phenomenon will be observed with the technological means necessary that won't leave a single doubt about its origin," General De Brouwer commented recently, with assurance. In the meantime, something physically, technologically real, yet completely unknown to any of us, repeatedly inserted itself into the skies over Belgium. We don't know where it was from, where it was going, or why it was there. But the fact of its existence is remarkable enough and a sufficient challenge to those of us below, unable to do a thing about it.

CHAPTER 2

The UAP Wave over Belgium

by Major General Wilfried De Brouwer (Ret.)

On November 29, 1989, when I was Head of Operations of the Belgian Air Staff, a total of 143 sightings were reported in a small area around Eupen, Belgium, thirty kilometers (nineteen miles) east of the city of Liège and eleven kilometers (seven miles) west of the German border. Some reported sightings were witnessed by more than one person, which means that at least 250 people described extraordinary UAP activity, with most reports occurring after sunset.

The weather was clear with open skies and good visibility. Two federal policemen, Heinrich Nicoll and Hubert Von Montigny, made the most important report. At 5:15 p.m., while patrolling on the road between Eupen and the German border, they saw a nearby field lit with such intensity that they could read the newspaper in their car. Hovering above the field was a triangular craft with three spotlights beaming down and a red flashing light in the center. Without making a sound, it moved slowly toward the German border for about two minutes and then suddenly turned back toward the city of Eupen. The policemen followed. Other independent witnesses reported that they saw the strange object along

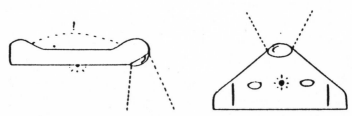

Drawing of sighting by two witnesses near Lake Gileppe, from the side and from underneath. SOBEPS archives

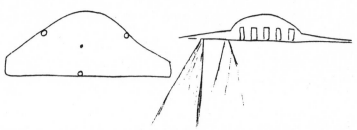

A witness in Eupen also drew the craft from two perspectives. SOBEPS archives

the same road. It remained over the town of Eupen for approximately thirty minutes and was seen by numerous additional witnesses.

The object then proceeded to Lake Gileppe, where it remained immobile, hovering for approximately one hour, while Nicoll and Von Montigny sat in their car on a nearby hill and witnessed an extraordinary spectacle. The craft repeatedly emitted two red light beams with a red ball at the spearhead of both beams, in the horizontal plane. Subsequently, the beams disappeared and the red balls returned to the vehicle. A few minutes later, another cycle started, each cycle lasting several minutes. Hubert Von Montigny said it was like a diver shooting an arrow from an underwater gun that slows down at the end of its trajectory and is subsequently retrieved by the diver.

But there was more to come. Suddenly, at 6:45 p.m., the policemen saw a second craft, which appeared from behind the woods and made a forward tilting maneuver, exposing the upper side of the fuselage. They described a dome on the upper structure with rectangular windows, lit on the inside. It then departed to the north. About forty minutes later, at 7:23 p.m., the first craft stopped emitting the red light balls and departed to the southwest. The two policemen, who were in radio contact with their dispatch, learned that another UAP had been reported in the north of Eupen, and they drove to an observation point, south of the highway E 40. From that position, they saw the UAP moving to the village of Henri-Chapelle, where two of their police colleagues, Dieter Plummans and Peter Nicoll (no relation to Heinrich Nicoll), saw the craft approaching from the direction of Eupen.

Plummans and Peter Nicoll stopped their car near a monastery, when they observed the craft with three very strong spotlights and a flashing red central light, at a distance of 100 meters (300 feet) and an estimated height of 80 meters (250 feet). The craft was immobile and silent, but it suddenly transmitted a hissing sound and reduced the intensity of the lights. Simultaneously, a red light ball came out of the center and headed straight downward, not far from their position.

The policemen were both terrified. The light ball turned from its vertical path into a horizontal path, and disappeared from view behind some trees. The craft moved then right above the police vehicle and headed to the northeast. They followed it for approximately five miles until they lost sight of it. Nevertheless, their colleagues Heinrich Nicoll and Hubert Von Montigny—the two policemen to first observe the objects a few hours earlier—could follow its movements from their position south of the highway.

In total, thirteen policemen reported seeing the craft at eight different locations in the vicinity of Eupen. Many civilians also saw the objects. For example, a family of four driving on a highway west of Liège saw a rectangular platform above them, made visible by the highway lights. They reported that it slowly passed overhead at a low altitude, with a spotlight in each corner.

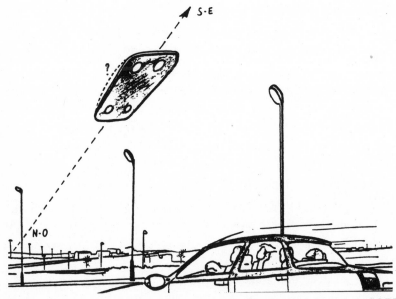

Sketch by a mother who observed a craft while on the highway with her family. SOBEPS archives

A total of seventy reported sightings made on November 29 were fully investigated and none of these sightings could be explained by conventional technology. Considering that approximately one person out of ten makes the effort to report their experience, the team of investigators and I estimate that more than 1,500 people must have seen the phenomenon at more than seventy locations from different angles during this afternoon and evening.

After the initial sightings on November 29, a series of sightings took place on December 1 (four observations) and December 11, 1989, when twenty-one witnesses reported similar descriptions of a triangular craft.

On December 1, air weather forecaster Francesco Valenzano and his young daughter, walking at the Square Nicolai in Ans, near Liège, saw a large slow-moving craft approaching at low altitude. The craft made a

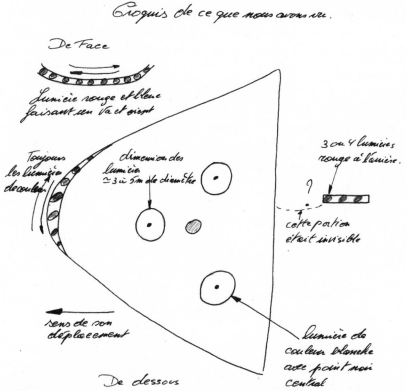

Valenzano's drawing included red and blue lights on the leading edge and four lights underneath. SOBEP archives

tour of the square without making any noise and when it passed directly over their heads, Valenzano noticed a delta shape with three lights in a triangular position and a red rotating light in the middle that was positioned lower than the belly of the craft.

On December 11, a twelve-year-old boy, along with his parents, grandparents, and sister, witnessed a similar-looking craft in the vicinity of their home for approximately fifteen minutes. It was at first immobile, and then started moving toward their house to pass vertically overhead. The boy's drawing shows a frontal view (bottom right), a view when it was almost overhead (bottom left), and a view when it was fully overhead (top). The different shapes could explain why some witnesses reported a craft that was not triangular. Indeed, the drawing shows that the perception of the shape can vary depending on observation angle and altitude.

About fifteen minutes later, a similar craft was observed approximately 97 kilometers (60 miles) more to the west, and several subsequent reports followed. At 6:45 p.m., Colonel André Amond, a civil engineer of the Belgian Army, was driving with his wife when they both saw three large light panels and a red flashing light at their right. He was driving faster than the craft, but when they stopped and got out of their car to observe the phenomenon, the light panels caught up and turned toward them. Suddenly, they saw a giant spotlight, about twice the size of the full moon, which approached them to an estimated distance of 100 meters. The colonel's wife was frightened and asked to leave. As he opened the car door, the craft made a very tight left-hand turn at a speed of approxi-

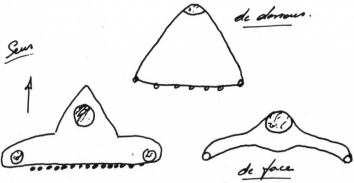

The craft from three angles, drawn by a boy in Trooze, near Liège. SOBEPS archives

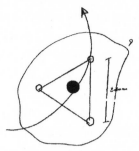

Colonel Amond sent his drawing of the UAP's ventral side to the Belgian minister of defense. A. Amond

mately 10 mph and three other lights appeared at the underside of the craft, in a triangular form with a central pulsing light.

There was no sound and, although it was a full moon, the witnesses didn't see the structure of the craft. After completing its turn, it suddenly accelerated very rapidly, only to vanish in the darkness of the night. Colonel Amond sent a detailed report to the Belgian defense minister. He ascertained that this craft was not a hologram, helicopter, military aircraft, balloon, motorized Ultra Light, or any other known aerial vehicle.

During a recent review of the investigation, it was learned that another witness had seen the object with three bright lights and a pulsating red light approximately five minutes before Amond and his wife. The exact timing could be reconstructed because she was walking home from the train that arrived at Ernage railway station twenty minutes before the Amonds spotted the craft.

On April 4, 1990, at 10:00 p.m. in the town of Petit-Rechain, a lady was walking her dog in her courtyard when she noticed the spotlights of a craft hovering above her home. She alerted her partner, who rushed outside with his newly bought camera. The camera was loaded with color slides, but only two shots remained on the film. Leaning against the wall to avoid instability, he took two photographs, the first with a manual exposure time of one to two seconds, while the craft was banking to the left. Subsequently, it started moving and disappeared out of sight behind the nearby houses. After the film was processed, the photographer saw four light spots on one slide and nothing on the second, which he threw away.

Several weeks later, he showed the remaining photograph to his fellow metalworkers during their lunch break in the factory. One of his friends contacted a local journalist, who published the photograph in a French magazine. From there, Belgian military academy experts were notified and requested the original slide for analysis. A team under the direction of Professor Marc Acheroy discovered that a triangular shape became visible when overexposing the slide.

After that, the original color slide was further analyzed by François Louange, specialist in satellite imagery with the French national space research center, CNES; Dr. Richard Haines, former senior scientist with NASA; and finally Professor André Marion, doctor in nuclear physics and professor at the University of Paris-Sud and also with CNES.

The major findings were:

- No effect of infrared radiation.
- No indication of any tampering with the slide.
- The camera was stable, but the craft was moving slowly and had approximately a 45-degree bank when the picture was taken.
- The rotation of the spotlights did not occur around one central point.
- The middle light is very different from the three other lights.
- The lights are positioned symmetrically with respect to the structure of the craft.

Professor Marion's more recent analysis in 2002 used more sophisticated technology. He confirmed the previous findings, while explaining a new discovery: Numeric treatment of the photograph revealed a halo of something lighter surrounding the craft. Special optical processing shows that within the halo, the light particles form a certain pattern around the craft like snowflakes in turbulence. This is very similar to the pattern of iron filings which is caused by "the lines of force" in a magnetic field. This could indicate that the craft is moving by using a magnetoplasmadynamic propulsion system as suggested by Professor Auguste Meessen in one of his studies.

Many hidden elements were revealed only through the analysis of this photograph, showing that the picture was not faked. The experts noted especially that the unique characteristics of the lights are very specific and said such an effect would not occur if the picture was a hoax. Also, the findings of the experts are consistent with the account of the photographer, who initially didn't think much of his shot of four strange lights and kept it in a drawer for weeks before showing it to anyone. He was not sure what it was, and for a while had not given it much thought.

Although the vast majority of the reports described a triangular craft with three spotlights and one flashing light at the bottom, as was captured in the Petit-Rechain photograph, a number of witnesses reported very special shapes and characteristics. On April 22, 1990, seven reports of triangles were submitted plus a more unusual report by two workers in Basècles, southwest of Brussels. They were in their factory courtyard shortly before midnight, when suddenly two enormous bright spotlights appeared, illuminating the courtyard. A huge trapezoid platform moved very slowly and silently slightly above the chimney, at one point covering the whole courtyard (100 x 60 meters, or 330 x 200 feet). The two men described six lights and said the color of the object was grayish. They saw structures at the bottom of the platform that looked to them like "an aircraft carrier turned upside down."

Another peculiar sighting, strikingly similar to the one at the Basècles factory, occurred on March 15, 1991, in Auderghem, near Brussels. An electronic engineer woke up during the night and heard a barely audible, high-frequency whistling tone. He looked out the window and saw a large rectangular craft at very low altitude with irregular structures

An artist's rendition of the "inverted aircraft carrier" at the Basècles factory. SOBEPS archives

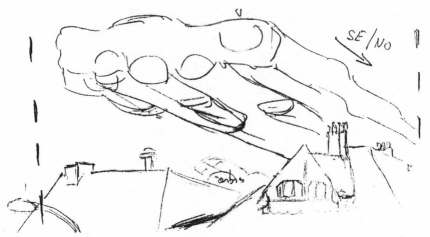

A witness's drawing of a rectangular craft over Auderghem, similar to the craft seen at Basècles a year prior. SOBEPS archives

on the bottom. Slipping on a jacket, he went upstairs to an upper-level terrace and watched the dark gray craft drift overhead very slowly without lights. The whistling tone had stopped and the craft was now silent.

A few days before, on March 12, 1991, a total of twenty-seven reports were filed from a small area southwest of Liège. On two occasions a craft was seen over the nuclear power plant of Thiange. One witness reported that it was directly above the red lights on the top of one of the enormous chimneys. It hovered there for approximately one minute, beaming one of its lights on the outside structure while another light pointed directly into one of the chimneys. After it had finished its "inspection," the UAP started moving slowly and flew straight through the enormous white plume of the chimney before disappearing in the dark.

Occasionally, a craft appeared to respond to the presence or actions of observers, as described earlier when Col. Amond stepped out of his car and the object immediately approached. On July 26, 1990, at 10:35 p.m., Mr. and Mrs. Marcel H. were also in their car, passing through Grâce-Hollogne and driving towards Seraing, when they looked out and saw an immobile object in the sky. It had the shape of an equilateral triangle that they estimated to be about twelve meters on each side. The object was dark, but a white-light belt, like a large neon tube, ran along two sides. The witnesses could see three spotlights beaming down toward the ground; they seemed to be detached from the object

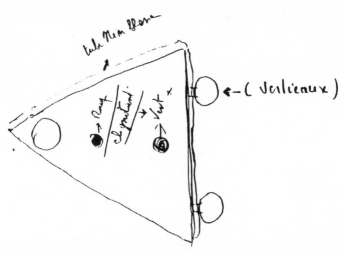

Drawing by Mr. Marcel H. He and his wife saw red and green flashing lights toward the center of the triangle, three large white lights, and a white neon tube. SOBEPS archives

but connected to each other by a sort of support "bracket." Also visible were two flashing lights, one red and one green, on the underside of the craft. The base—the side with two white spotlights—was facing toward them.

Surprised, Mr. H. said to his wife: "For the fun of it, I am going to flash my lights." Mr. H. flashed his car lights twice—off and on, off and on. At this same moment, the two white lights at the base of the triangle rotated, tilted toward the two passengers below, and flashed off and on three times. The illumination was bright, but not blinding. Then, keeping the lights pointed at the moving car, the object proceeded toward the vehicle and, moving with the base forward, positioned itself to the right at a distance of approximately 100 meters and a height between 60 and 100 meters. (It is interesting that Col. Amond also reported a distance of 100 meters after the object approached.) It then made a banking turn and, still moving with its base forward, flew in the same direction as the car, following it as it continued its downhill course toward Seraing. Although the hill was rather steep, the UFO moved with the terrain and maintained a constant height above the sloping ground, flying at the same speed that the car was driving (60–70 km/hr). By the time they approached the bridge at Seraing, Mr. and Mrs. H. were quite frightened. Finally, the object crossed the river Meuse right next to them without making any noise, and then started to climb, rapidly departing in the direction of Grâce-Hollogne.

A lengthy book could be written with nothing but witness reports and drawings collected during the two peak years. I have presented only a sample. I can conclude with confidence that the observations during what is now known as the Belgian wave were not caused by mass hysteria. The witnesses interviewed by the investigators were sincere and honest. They did not previously know each other. Most were very surprised by what they saw, and today, twenty years later, they are still prepared to confirm their unusual experience. Those close to the craft were frightened or terrified; one fell off his bike and was in shock. Several witnesses had high-ranking functions and preferred not to reveal their names to the media.

Of the approximately 2,000 reported cases registered during the Belgian wave, 650 were investigated and more than 500 of them remain unexplained. It is logical to assume that many thousands more witnessed UAP activities and did not report them. The findings were exceptional. More than 300 cases involved witnesses seeing a craft at less than 300 meters (1,000 feet), and over 200 sightings lasted longer than five minutes. Sometimes observers were right underneath the craft.

Although many questions remain unanswered, analysis shows that a number of points can be made with certainty, and some conclusions can be drawn.

- Most witnesses reported the craft had a triangular shape, but a number of reports mentioned other shapes, such as a diamond, cigar, or egg, and, in a few spectacular cases, an aircraft carrier turned upside down.
- The reported air activities were unauthorized, yet were observed by multiple witnesses while not registering on surveillance radars.
- It can be deduced that on both November 29 and December 11 at least two crafts were active at the same time. On November 29, two policemen reported two at the same time in different locations, and also different shapes were reported. On December 11, witnesses reported seeing a craft at the same time at different locations.

- On several occasions, the craft made a tilting maneuver allowing observers to see its upper side, revealing a dome at the top. Some reported windows or lights on the side of the craft; others saw lit windows in the dome.
- No electromagnetic effects, such as radio interference, were experienced.
- Not one aggressive or hostile act was noted.
- The flying objects didn't try to hide and, in several cases, moved toward the observers on the ground. Some witnesses reported that crafts responded to their signals, such as switching one of its lights off and on when they flashed the headlights of their car.
- The crafts performed in ways not possible by known technology. They were able to remain stationary and hover, even in unusual positions such as vertical and/or banking at 45 degrees or more. They could fly at slow speeds and accelerate extremely fast, faster than any known aircraft, and they remained silent, or made only a very slight noise, even when hovering or accelerating. The objects were equipped with very large spotlights, with a diameter of more than one meter (three feet), capable of intensively illuminating the ground from an altitude of 100 meters (330 feet) or more. The integrity of these lights was variable—in some cases, witnesses reported that the lights were not illuminating the ground and were not blinding. Experts are convinced that the spotlights are of a very special nature; the size and intensity have not been seen on any aircraft before. These crafts carried a red light from underneath, at the bottom of their underside and apparently unattached to the structure, which seemed to pulsate rather than rotate. On three occasions, red light balls left the structure, and on two occasions they were seen returning to the craft.
- Some of these individual performance capabilities may be explainable in isolation, but the combination of all of them makes them highly unusual, even enigmatic. The technology used by these crafts was so advanced that even today, twenty years later, it is not available.
- The most important conclusion is that there must have been air activities of unknown origin in the airspace of Belgium. The number of cases and the credibility of the vast number of witnesses leave us with an intriguing mystery.

The events of November 29 were extensively covered by the media and naturally the Air Force was overwhelmed with questions. The questions were addressed to the Belgian defense minister but ended up on my desk as Chief of Operations of the Air Staff. I was asked repeatedly about the origin and nature of these craft.

The Belgian Air Force tried to identify the alleged intruder(s). We verified the radar registrations for November 29 but nothing special had been recorded. In addition, the civil aviation authorities confirmed that no flight plans had been introduced and that no special activities had been recorded by the civil radars. I was able to determine that the objects seen on November 29 could not have been helicopters, blimps, or any fixed-wing aircraft. This implied that the reported object(s) committed an infraction against the existing aviation rules.

We were dealing with a problem. I checked further to find out if these objects could have been espionage flights made by F-117 stealth aircraft or anything similar. Because of the described performances that did not match any known technological capabilities, I was convinced that this was not the case. I also couldn't believe that any other nation would conduct experiments with crafts using unknown technology over a populated area without any formal authorization. Nevertheless, I forwarded the question to the U.S. Embassy, which quickly confirmed that no Stealth flights or any other experimental flights had taken place over Belgium.

Because there was no explanation for the events of November 29, and also because the sightings continued, we agreed to authorize the national defense system to scramble two F-16 fighter jets when abnormal activities were reported. The first two F-16s were sent out on December 8 after strange lights were reported, but nothing definitive was determined.

In cooperation with the civil aviation authorities and the federal police, the Air Force established a procedure by which the F-16s could identify these phenomena. To ensure that the fighter jets would not be scrambled irresponsibly, we decided that authorization to launch the Quick Reaction Alert (QRA) aircraft would only be given when: (1) the sighting of a craft was confirmed by the police, and (2) the object was detected on radar. This meant that the radar stations had to pay special attention to slow-moving targets when notified of an observation by the police.

This would avoid unnecessary scrambles, but it also had major disadvantages. Most of the witnesses didn't react by calling the police, or weren't able to call quickly enough—mobile phones didn't exist yet—for the police to confirm the sightings. It was also problematic for radar controllers to work on a screen that was heavily cluttered, in order to be able to record targets not usually shown on the scope. Thus the precautionary measures prevented quick scrambles.

As Chief of Operations at the Air Staff, I felt obliged to closely follow the events. However, no priority was given to this by the Belgian government since no threatening incidents had occurred, and no formal inquiries were conducted by any governmental bodies. Although the defense minister insisted on a transparent approach, especially to show the public that there was no cover-up, the Air Force was not authorized to establish a dedicated office for conducting its own inquiries. Instead, the Air Force supported SOBEPS—the scientific research group investigating the case—in any way we could, such as providing information on registered air activities over observation areas and responding to requests for radar data. SOBEPS approached the issue professionally and the Air Force information allowed the organization to make objective inquiries and file all relevant data.

On the evening of March 30–31, 1990, an F-16 launch was initiated after the observation of strange lights by several policemen, and after an assumed flying object was confirmed by two military radar stations. Once aloft, the pilots tried to intercept the alleged crafts, and at one point recorded targets on their radar with unusual behavior, such as jumping huge distances in seconds and accelerating beyond human capacity. Unfortunately, they could not establish visual contact.

The defense minister received many follow-up questions about this launch, but the Air Force needed time to properly analyze the data. We called a press conference about three months later, on July 11, 1990. The activities of the F-16s had been reconstructed, but the technical analysis was not fully completed. I presented one particular radar lock-on that showed extraordinary accelerations well outside the performance envelope of any known aircraft. Nevertheless, I added that this needed further analysis by experts because these types of returns could have been caused by electromagnetic interference.

It turned out that only one F-16 camera had made satisfactory radar recordings, so comparison between the aircraft recordings was not possible. This was a serious problem. A comparison would have allowed us to exclude those returns that were caused by electromagnetic interference, because the data from such interference are never the same on two different radars. Therefore, we couldn't be sure if the radar echoes were caused by electromagnetic interference or by something unusual.

The conclusion of the Air Force, therefore, was that the evidence was insufficient to prove that there were real crafts in the air on that occasion.

The Air Force's decision that the evidence was insufficient to conclude that there were unusual air activities during the night of March 30, 1990, was gleefully accepted by the irrational skeptics and the debunkers, who immediately claimed that that whole Belgian UAP wave was a farce. For them, one explainable case is enough to discredit the more than five hundred remaining unexplainable sightings—a position that is still put forward by most of them today.

In 1990, the Air Force stated on several occasions that it had no explanation for the numerous sightings. Today, persistent skeptics, who make a point of publicizing their positions, have come forward with a theory that these were helicopters. At the time of the UAP wave, the Belgian Air Force was working with civil aviation authorities and had more than 300 aircraft—including helicopters—several ground radar stations, 500 pilots, more than 300 engineers, 100 controllers, and thousands of technicians, etc., but we were not able to find the answer.

Even so, a few unqualified debunkers claim to have found that answer. Their real objective is to misinform people, create confusion, and ridicule UAP sightings. Some witnesses who made reports in 1989 are still hounded and discredited to this day. No wonder that several witnesses didn't dare to reveal their names; some didn't even take the risk of reporting their sightings. I had personal experience with two different people—a journalist and a NATO employee—whom I knew for many years: They verbally reported two sensational sightings but didn't want (or dare) to put anything on paper.

The approach to the UAP problem has to be critical but objective. Indeed, we are dealing with a very important question: Is our airspace being violated by unknown intruders? False claims and disinformation by people trying to ridicule the UAP phenomenon are made use of by those who refuse to accept that some sightings remain unexplainable and could possibly be some kind of unknown technology. Sadly, this not only strongly impacts the witnesses, but it also diminishes the sense of responsibility within government. None of our political leaders want to be involved in UAP issues. Knowing that the vast majority of the population is more concerned with their immediate and short-term needs, political leaders focus on resolving these problems as opposed to longer-term, strategic issues. They avoid any connection with UAP because they are afraid of being ridiculed and losing credibility with the public. It is perceived to be like a hot potato—don't touch it or you will burn your fingers.

The vast majority of military leaders almost automatically reject any responsibility for investigating UAP sightings, because this does not figure in their terms of reference. They devote all their time and energy to the ongoing operations and don't feel concerned about issues on which they have no firm grip. In addition, if not experiencing any direct threat from unidentified aerial phenomena—to my knowledge, no recent security incidents have been reported—investigations of UAP sightings are not on the priority list of military commanders, and they will not initiate investigations. UAP reports are considered as a hindrance, a time-consuming interference with normal routines.

One easy way for the authorities to stop the flow of annoying questions is to give a false explanation for reported phenomena, as has been done many times. To a certain extent, such tactics work to hush up the hype, in particular if there is only one event. But this does not deal with the substance of the problem. On the contrary, it creates an atmosphere of distrust and suspicion between those who witnessed the event and the responsible authorities.

For the military, it becomes more problematic when the events occur not just once but multiple times. The defense authorities are under pressure to provide an acceptable answer. Unfortunately, during the Belgian UAP wave, no such answer could be found.

There is only one solution and this is to tell the truth. The truth is that the Air Force could not determine the origin of the objects witnessed by thousands of people. It is not easy to admit that authorities in charge of air defense and airspace management are not capable of finding an acceptable explanation, but, in my opinion, this is better than issuing false explanations. The Belgian government was honest and acknowledged publicly that it could not explain the many sightings.

Nevertheless, military authorities should not wait to take action until they are forced to do so by the public and the media. They should be concerned about the possible security implications of unusual air activities. If reliable witnesses report the presence of UAP that have not been picked up or identified by the civil aviation authorities and the air defense systems, it should be admitted that there may be a problem and an effort should be made to conduct more in-depth investigations with qualified experts.

What if these crafts had more aggressive intentions? Who would have been responsible if incidents had occurred? The question remains: Which military authority dares to tackle this problem, or rather, which military authority dares to recognize that there is a problem? Is this "ostrich" policy the right approach?

Formally investigating reliable UFO reports would create an atmosphere of openness and transparency, and motivate other witnesses to come forward with their experiences. Such investigations would provide the scientific basis for relevant authorities to express an official opinion vis-à-vis the UFO problem. However, it seems a wake-up call will be required for us to formally acknowledge that there is a problem. A major accident would serve as such a wake-up call, but this is not what we hope for; on the contrary, this is something that we want to prevent. We must all be prepared for the next UFO wave, wherever it may occur.

CHAPTER 3

Pilots: A Unique Window into the Unknown

The Belgian UFOs did not appear to create any kind of safety hazard for aircraft in flight, as far as we know, and General De Brouwer made it clear that the objects displayed no threatening behavior. Yet, as I stated in the second point to be considered in the Introduction, this is not always the case. Some of our most compelling reports on UFO encounters have been provided by Air Force and commercial pilots, and sometimes aviation safety is compromised.

Shortly after publishing my first story about the COMETA Report in the *Boston Globe*, I became interested in the question of UFOs and aviation safety. After all, if these things really are out there, one would expect that at least *some* pilots would see dazzling light displays while flying at night, or maybe giant triangles in the daytime, or metallic discs speeding by the cockpit window. In fact, wouldn't they be more likely to see them than anyone else? Perhaps passengers might even be at risk if they found themselves too close to an unpredictable unidentified flying object. One could easily imagine that witnessing such a thing at 35,000 feet—something with no wings but much faster and more agile than the lumbering jet aircraft holding one prisoner—must be considerably more unnerving than viewing the same object with one's feet safely planted on the ground. But beyond simply seeing one, could they be dangerous?

Much to my amazement, I quickly discovered that a ninety-page report dealing with this very question had just been released by the world's most qualified researcher of pilot encounters with UFOs. Even better, I recognized that this well-documented scientific study could serve as the "news hook" for another story, in the same way that the COMETA Report had done before. "Aviation Safety in America—A Previously Neglected Factor" by Dr. Richard Haines, a retired senior research scientist from NASA Ames Research Center and former chief of NASA's Space Human

Factors Branch, was a mind-boggling study, with more than fifty pages of case summaries involving pilots and their crews. That "neglected factor," of course, referred to unidentified aerial phenomena, or UAP.

The report featured over one hundred cases of pilot encounters with a variety of these UAP, including fifty-six near misses, all affecting the safety of aircraft. Most cases involved multiple witnesses, and many were backed by ground radio communications and radar corroboration. Experienced pilots presented accounts of objects, ranging from silver discs to green fireballs, flying loops around passenger aircraft, pacing alongside despite pilots' evasive attempts, or flooding cockpits with blinding light. Dr. Haines documented cases of electromagnetic effects on aircraft navigation and operating systems linked to nearby UFOs, or a pilot's sudden dive to avoid a collision. He wrote that a crew's ability to perform its duties safely is disrupted when crew members are faced with "extremely bizarre, unexpected and prolonged luminous and/or solid phenomena cavorting near their aircraft." The danger posed by the phenomenon in flight lies more with the human response to it than from the actions of the UAP itself, because the objects do not appear to be aggressive or hostile, and seem to be able to avoid collisions by executing last-minute high-speed turns in a flash.

Dr. Haines, who has authored more than seventy papers in leading scientific journals and published over twenty-five U.S. government reports for NASA, specialized in human performance, technology design, and human-computer interaction while at NASA. Having contributed to the U.S. Gemini and Apollo projects, as well as Skylab and Space Station, in 1988 he retired from his twenty-one years as a senior aerospace scientist at NASA Ames Research Center. Subsequently he worked as a senior research scientist for the Research Institute for Advanced Computer Science, RECOM Technologies, Inc., and Raytheon Corp. at NASA Ames Research Center until 2001.

Haines unexpectedly became interested in the UFO subject back in the 1960s, when he was conducting research involving flight simulators for NASA. As he explains it, commercial pilots would volunteer to come into his facility and fly the simulators for studies on aviation safety, avionics, and many other areas. "From time to time a pilot would offer to tell me about an experience he had that just blew me away," Haines said in a 2009 interview. Although he had heard of UFOs at the time, he had absolutely no interest in them. "I heard more and more stories from these very credible witnesses, so it began to catch my attention. I said to myself, 'I can explain these things; they're all natural phenomena or misidentified phenomena within the human eye,' which I knew a lot about from studying human vision and optics. So I set out as a skeptic to disprove the whole thing. But the more I looked into the subject seriously, the more convinced I became that there was something there. Something that deserved to be looked at. Yet none of my colleagues were doing so." He then started systematically collecting data and eyewitness reports, and giving a great deal of thought to the analysis, and has been doing so ever since. Today, he has developed an international database of over 3,400 firsthand UFO sightings by commercial, military, and private pilots, with special attention to cases where aviation safety is compromised, as distinct from sightings during which the objects have no effect on an aircraft or its crew.

In fact, for years, he and his associates have been attempting to alert the aviation community to the effects of unknown aerial phenomena on aircraft safety. In 2001, along with executive director Ted Roe, he established the National Aviation Reporting Center on Anomalous Phenomena (NARCAP), a respected international nonprofit research organization serving also as a confidential reporting center for use by pilots, crew, and air traffic controllers who are otherwise afraid to make reports of sightings. NARCAP scientists collect and analyze high-quality data to further understand the fundamental nature of all kinds of unidentified aerial phenomena that may pose a threat to aviation safety. The group's technical and science advisors with extensive aviation and aeronautic experience from about a dozen countries, along with other specialists ranging from geophysicists and research psychologists to meteorologists and astrophysicists, contribute research and publish "Technical Reports" on the group's website.

I have been privileged to come to know Dr. Haines, and he invited me to sit in on a number of private NARCAP annual meetings over the years, the last one being in July 2008. I was honored to meet many of these dedicated professionals, who are doing an outstanding job despite the obstacles they face. Papers and ongoing research are presented at these round-table gatherings, and strategies are discussed for acquiring greater accessibility to the aviation community, making sure that

NARCAP remains distinct from activist UFO groups where aviation safety is obviously not the focus and where a rigorous scientific approach is less often employed.

Nonetheless, the group's efforts to bring this issue into the scientific arena and aviation community have fallen on deaf ears. "There is little doubt in my mind that no amount of rational discussion about the substantiated evidence of the presence and behavior of UAP in our skies is going to quickly overcome the impact on two generations of Americans repeatedly told otherwise: that the subject of UAP should, at best, be cast into the category of folklore and, at worst, viewed as somehow harmful propaganda," Dr. Haines commented recently in an e-mail. "But we must keep working toward the goal of accepting the truth when and where we find it. To do anything less is to set ourselves up for a possibly dangerous future."

Beyond the legitimate efforts to confront safety issues, I became intrigued by the absolutely crucial and central role pilots can play in simply documenting these mysterious and elusive UFOs, whether safety is a factor or not, since they represent the world's most experienced and best-trained observers of everything that flies. Able to rapidly identify and respond to anything that would endanger a flight, pilots are required to have practical knowledge of all other aircraft, military test flights, and other special air activities such as missile tests, as well as unusual weather and natural phenomena. Professional pilots are highly qualified to recognize a true anomaly as distinct from any of these. What better source for data on UFOs is there? The aviation world is in a position to provide information that could greatly increase knowledge about the UFO phenomenon, if only our scientists wanted to take advantage of it.

These professionals spend countless hours behind a unique window into miles of usually empty sky, a perfect platform for observing exceptional details about the behavior and physical appearance of UFOs when they appear. Pilots might be able to precisely determine the distance and velocity of the anomaly, as well as its relative size, which is more difficult to estimate from the ground. They could also document the transitory impact of electromagnetic fields on cockpit equipment, providing potentially useful clues as to the nature of any radiation from the object. Able to remain calm and focused during unexpected stressful situations, pilots can report accurately and precisely on events outside, using on-board radar and communications with air traffic control with its ground radars to home in on the object. Nearby aircraft could be contacted and asked to head for the area, or military jets could be launched if the encounter was prolonged. And—of great interest to all of us—crew members would be able to take outstanding photographs and videos of the lengthier encounters. These unique circumstances potentially transform any jet aircraft into a specialized flying laboratory for the study of rare anomalous phenomena. Important evidence of UFOs has been obtained this way in many powerful cases since the 1950s, not only raising concerns about safety, but also adding greatly to the historical record.

Pilots are among the least likely of any group of witnesses to fabricate or exaggerate reports of strange sightings. But unfortunately, as things stand now, most would prefer never to be confronted with the dilemma of seeing a UFO and having to decide whether to report it. According to Haines, reporting on the presence of UAP *has* been enough to threaten some pilots' careers, and for this reason, most choose not to do so.

Neil Daniels, a United Airlines captain for thirty-five years, with more than 30,000 hours of flying time and an Air Force Distinguished Flying Cross, was one of those pilots who feared reporting his sighting, despite the physical effect experienced by his airplane. In 1977, he, his copilot, and a flight engineer observed a perfectly round, "brilliant, brilliant light off the wing tip," as he described it, about 1,000 yards away from their United DC-10, which was en route to Boston Logan from San Francisco. While flying on autopilot, the passenger plane was forced into an uncommanded left turn, apparently pulled by the object's magnetic interference, prompting Boston Center to ask, "United 94, where are you going?" Captain Daniels replied, "Well, let me figure this out. I'll let you know."

The captain and his first officer then noticed that their three compasses were all reading different headings, and at that point they deliberately uncoupled the autopilot and flew the airplane manually. (Haines points out that the magnetic sensor providing the input to the compass then controlling the autopilot was the one located nearest to the UAP.) The powerful light followed along with the aircraft at the same altitude for several minutes, and then took off rapidly and disappeared.

Captain Daniels said that the luminous object shot away so swiftly that he does not understand how it could possibly be a man-made machine. But no matter what it was, he says, "it did cause a disruption in the magnetic field around the aircraft to the point where it pulled the aircraft off course."

Neither Daniels nor any of his crew reported the incident. The air traffic controllers did not ask further questions about the disturbance to his flight. It was as if everyone wanted to pretend that nothing had happened, but Daniels could not forget what he had seen with his own eyes. Seven months later, while duck hunting with his United Airlines boss, he had a momentary change of heart and decided to tell him the story. Unfortunately, he discovered that his initial instinct to keep quiet was the right one. "I'm sorry to hear that," Daniels's employer admonished. "Bad things can happen to pilots who say they have these sightings."

Now retired, Daniels was not particularly concerned about the safety of his jet at the time. But if, as Daniels reported, a UFO can knock a flight off course from a distance, what might happen if it were closer?

CHAPTER 4

Circled by a UFO
by Captain Júlio Miguel Guerra

In 1982, Portuguese Air Force pilot Júlio Guerra happened to look from his cockpit window down toward the ground below, and saw a low-flying metallic disc. Suddenly, it bolted up toward him at high speed. During a lengthy series of events, this object demonstrated a harrowing variety of maneuvers at close proximity to Guerra's small plane, witnessed by two other Air Force pilots called to the scene. Since leaving the Air Force in 1990 after eighteen years of service, Guerra has been a captain with Portugália Airlines, Portugal's largest commercial airline. He's never seen another UFO, but remembers this life-changing event with tremendous clarity.

On the morning of November 2, 1982, I was flying a DHC-1 Chipmunk northward in the region of Montejunto mountain and Torres Vedras near Ota air base. It was a beautiful, clear day with no clouds, and I was headed in the direction of my work area, E (echo) zone, planning to climb to 6,000 feet for an aerobatic training. As a twenty-nine-year-old lieutenant with ten years in the Air Force, I was a flight instructor as part of 101 Air Force squadron, flying solo in my plane.

At about 10:50 a.m., when I was overflying Maxial zone at an altitude of 5,000 to 5,500 feet, I noticed below me and to the left, near the ground, another "aircraft." But after a few seconds I saw that this airplane seemed to have only a fuselage. It didn't have wings and it didn't have a tail, only a cockpit! It was an oval shape. What kind of airplane could that be?

I immediately turned my plane 180 degrees to the left in order to follow and identify this "object," which was flying to the south. Suddenly

47

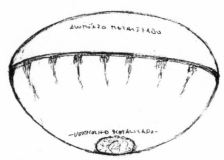

I made this drawing the day after the encounter and submitted it to the Portuguese Air Force. On top I wrote "metallic aluminum" and underneath "metallic red." J. Guerra, CNIFO Case Report

the object climbed straight up to my altitude of 5,000 feet in under ten seconds. It stopped right in front of me, at first with some instability, oscillations, and a wavering motion, and then it stabilized and was still—a metallic disc composed of two halves, one on the top and another on the bottom, with some kind of band around the center, brilliant, with the top reflecting the sun. The bottom half was a darker tone.

At first it moved with my aircraft, then it flew at a fantastic speed in a large elliptical orbit to the left, between 5,000 feet to the south and approximately 10,000 feet to the north, always from left to right, repeating this route over and over. I tried to keep it in sight.

Right away, when I realized it was an unknown object, I called the tower and told the controller that there was a strange object flying around me. He, and others from three or four other airplanes, said it must be some kind of balloon. Some of the pilots flying in other zones made fun of it, and I responded by asking them to come and see it with their own eyes if they didn't believe me. I told them that if it was a balloon, how could it ascend from the ground to 5,000 feet in a few seconds? The response was silence. They started asking for my location, my work zone, and two fellow Air Force officers, Carlos Garcês and António Gomes, told me they would join me.

While waiting and watching it, I wanted to know more about this object. Even though I got close, I didn't know what it was. I was alone with it for fifteen minutes—which felt like forever—never knowing what would happen next or if it would come back each time it set out on its course. I stayed there and focused on this thing repeating its elliptical course around my aircraft.

When Garcês and Gomes arrived in their Chipmunk after about fifteen minutes, they radioed "Where is it?" I gave them the position, and after they saw it I felt better, because now two more Air Force pilots had seen the same thing I did. They stayed with me for about ten minutes while the object kept up its circular pattern, each loop almost the same as the previous one, and we conversed on the radio. I was in the interior of the orbit and they were outside of it, so the object passed between the two planes. Because of that, we could estimate the size relative to the length of the Chipmunk's fuselage (7.75 meters): about eight to ten feet.

After about ten minutes, I still was curious and really wanted to know more about this object, so I decided to make an interception, meaning I would head directly toward it but slightly to the side, so it might be forced to alter its course. I told my two Air Force colleagues there that I was planning an intercept. Since the object's speed was much faster than my own, I flew directly to a point along the trajectory of its elliptical course. It came toward me and flew right over me, on top of my aircraft, and stopped there, like a helicopter landing but much, much faster, breaking all the rules of aerodynamics. It was very close to my plane, only about fifteen feet. I was astonished. I closed my eyes and I froze at this moment, without reacting.

There was no impact . . .

It then flew off in a flash toward the direction of Sintra mountain, to the sea. All this happened so fast that I couldn't do anything with my aircraft to try to avoid the object. One of the other pilots saw the whole thing.

At various times the object had been very close to me and I was able to verify that it was round with two halves shaped like two tight-fitting skullcaps. I carefully looked at the lower one, which seemed to be somewhere between red and brown with a hole or dark spot in the center. The center band looked like it had some kind of a grid, and possibly a few lights, but it was hard to tell since the sun was so bright and was reflected.

Right after landing, all three of us filed detailed, independent written standard reports about the incident, and our planes were checked for damage, but we didn't hear anything more about it from anyone in the Air Force, and we were not interviewed by the military. A little later, General José Lemos Ferreira, the Portuguese Air Force Chief of Staff at the time, authorized the release of all the records to a team of scientists and experts.

In 1957, General Ferreira had witnessed an unknown luminous object himself when he was leading a nocturnal flight between Ota air base, Portugal, and Cordova, Spain. Three other Air Force pilots flying in separate planes saw the phenomena as well—at first one large object and then four small "satellites" that came out of it. He was aware of the scientific importance of these types of things, and he sent a report on this incident to Project Blue Book, run by the U.S. Air Force.

Since the general had some understanding about UFOs, he released all information held by the Portuguese Air Force about my encounter, and a lengthy scientific investigation was launched in 1983 and completed in 1984. The team of experts included about thirty people from different disciplines and academic institutions, including historians, psychologists, physicists, meteorologists, engineers, and other scientists.

This investigation involved cooperation between the military and the civilian scientists. I went back to the zone and flew in the path that the object took for its initial vertical ascent at the time I first saw it, which took only a few seconds. Estimating ten seconds, and covering the same distance, we determined that it was flying at over 300 mph vertically. This is not possible for a helicopter, and, more importantly, a human being inside could not survive the g-force from the acceleration required for this upward motion.

Since I could show the investigators on the map the trajectory of the object in its elliptical orbits relative to points on the ground, they could determine its velocity to be about 1,550 mph. This speed is incredible, especially given the maneuvers it was making. I don't know if it was from another universe or planet, or if it was from the ground here; I simply don't know. I have never seen anything else like this since.

The scientific team studied all the data and the three pilot reports, and after a meeting of all thirty investigators in Porto, in 1984, the group provided a written analysis of more than 170 pages. They did everything they could to understand this case, but they could not find an explanation for it. They concluded that the object remained unidentified.

I talked about my experience to the media and had no problems; it was covered seriously in many newspapers and on TV because we had three Air Force pilots involved. Since then, people have approached me to tell me about other UFO incidents, but most of them want to keep their experiences private.

Another incident occurred in Portugal before mine. An Air Force pilot, a colleague, saw a portion of an object behind clouds, which appeared to have two or three windows. He lost control of his Dornier Do 27 aircraft; it began to fall, and he only regained control right above the trees. His comments were on the air traffic controller tape, and he thought this was the end. I was there at the base when he landed and he talked with a group of us right away and filed a report. The engineers tried to figure out how he lost control. Afterward, some engineers outside of the Air Force came to the hangar where his plane was parked with many other identical aircraft. They were able to locate his particular plane by using an instrument that measures radiation; his plane registered high, and this could not be explained.

This pilot went on to have a career as a civilian pilot, as I did. After eighteen years in the Air Force, ending in 1990, I began flying commercially and am now a captain with Portugália Airlines (TAP), though I still fly solo. I still don't know what I saw that day back in 1982, but I love flying as much now as ever. My encounter, though incredible, did nothing to change that.

CHAPTER 5

Unidentified Aerial Phenomena
and Aviation Safety
by Richard F. Haines, Ph.D.

Safety is of prime importance to everyone who flies or is associated with flying. Yet most Americans have never heard about the fact that UAP sightings can affect flight safety. These incidents are not investigated by any government agency, as are other events affecting aircraft. In fact, aviation officials have prevented them from coming to light by censoring the reporting process in various ways. Pilots who make an official UAP report continue to be ridiculed by government officials and/ or their own airlines and instructed not to report their sighting publicly. This attitude serves no one and in fact puts all of us at greater risk while traveling in airplanes. It prevents the scientific community from acquiring the data necessary to investigate the origin of these UAP, and it also keeps airlines and pilot organizations from taking action or providing their pilots with specialized training and safety protocols. Despite all this, these unusual aerial phenomena have continued to plague commercial, military, and private flight operations over many years.

The near-miss incident described by Lieutenant Guerra over Portugal in 1982 provides a powerful example of a case in which aviation safety was challenged by an unidentified object, according to virtually any standards: military, private, or commercial. Whenever another airborne vehicle of any kind cannot be communicated with, makes a very high-speed approach, and then stops unexpectedly within fifteen feet of one's airplane, any pilot in the world would be justifiably concerned and even afraid. Lieutenant Guerra and his two fellow pilots are to be commended for reporting this bizarre incident to officials, although the pressures against doing so are less intense in Europe and South America than they are here. In addition, General Ferreira, the Portuguese Air Force Chief of Staff at the time, willingly made all the records available to a scientific research group qualified to investigate—a scenario that we unfortunately do not see in the United States. Yet all countries are equally affected by the fact that UAP can appear without warning at any time and place.

Three kinds of UAP dynamic behaviors and their consequences have been consistently reported. First and foremost are near-miss and other high-speed maneuvers by UAP near airplanes. Many cases involve a relatively small distance—generally on the order of tens of yards—between the aircraft and the reported aerial phenomenon, which qualifies them as near misses by federal aviation standards in the United States and the United Kingdom. While a pilot's estimate of the distance to a UAP may be affected by darkness or the lack of reliable visual distance cues, these highly trained professionals are generally quite accurate and usually will not be in error by more than an order of magnitude.

Fortunately, the immediate physical threat of in-flight collision seems unlikely because of the high degree of maneuverability exhibited by the UAP. In many cases, the objects rapidly avoid a collision at the last minute, and it's not left up to the pilots to make these moves. But in some cases, pilot reactions can be a problem, as well. In order to avoid a perceived collision with UAP, some have made violent control inputs resulting in passenger and flight crew injury. And there is always the danger that if a pilot makes the wrong control input at the wrong time during an extremely close encounter, a midair collision could occur.

In one example, a U.S. Air Force Boeing KB-50 aerial refueling tanker was making a night landing at Pope Air Force Base in North Carolina when the pilot and crew noticed an object and saw strange lights. On their final approach, the pilot had to maneuver around the object and climb again to wait for it to depart. Air Force tower personnel saw the UAP hovering above the airport, and watched it through binoculars for twenty minutes, stating that it was not an atmospheric phenomenon of any kind. Air Force officials acknowledged that "the UFO presented a hazard to aircraft operating in the area"—one of the few official statements to this effect on record.

The second impact that UAP can have on aviation safety is to affect the proper functioning of navigation guidance equipment, flight control systems, radar operations, and radio communication with interference from its alleged electromagnetic radiation. Obviously, in situations where pilots rely on their instruments, the probability of an incident or accident increases when anomalous electromagnetic effects cause them to malfunction. Fortunately, in most of these instances, equipment resumes normal functioning after the object departs.

Finally, cockpit distractions produced by close encounters with UAP divert the attention of the crew and can impair their ability to fly the airplane safely. It is understandable that witnessing bizarre objects or unexplained lights pacing beside an airplane, or flying circles around it, would be disconcerting to anyone on board, especially those responsible for passenger safety.

The information I've collected to document cases of UAP affecting aviation safety comes from my extensive database. It consists of pilot and air traffic control reports drawn from official U.S. and other government sources, private interviews, and reports by international colleagues who have worked closely with the National Aviation Reporting Center on Anomalous Phenomena (NARCAP). According to our statistics, in an average career of commercial flying, a pilot has about the same chance of seeing a UAP as he does of striking a bird in flight or of encountering extreme wind shear. The threat to safety is small but potentially significant, and should be treated like any other infrequent safety hazard. Many flight safety problems go unreported or underreported, but the difference here is that bird strikes and wind shear are currently acceptable events to report and UAP are not.

Three cases over Australia and New Zealand are of great interest, illustrating the effects I'm referring to. On August 22, 1968, at about 5:40 p.m., two pilots were flying from Adelaide to Perth, Australia, at 8,000 feet in a Piper Navajo single-engine airplane when they sighted a very large cigar-shaped object surrounded by five smaller objects. The strange formation maintained a constant angle from their own flight path for over ten minutes, while they flew at 195 knots. One of the pilots said later, "The large one opened up in its center with smaller objects going to and from the larger object." Ground air traffic control was contacted and replied that there was no known air traffic in the area. At this point their radio failed on all frequencies until the objects flew away, "as if by a single command."

Ten years later, a shocking event occurred. A private pilot went missing while en route to King Island, south of Melbourne, Australia, after

a very close and frightening encounter with a large unknown object.
On October 21, 1978, twenty-year-old Frederick Valentich had rented a
Cessna 182L single-engine, propeller-driven airplane for a short night
flight. Just after 9:00 p.m., he radioed Tullamarine Airport in Melbourne
from an altitude of 4,500 feet while over the waters of Bass Strait. For
six and a half minutes, he conversed with flight service specialist Steve
Robey at the Melbourne airport about something unidentified orbiting
around his airplane, heading straight for him, and chasing him. The
tape ended with fourteen seconds of very unusual metallic noises and
then went silent.

The voice transcript between Robey at Flight Service in Melbourne
and Valentich in his Cessna aircraft—which was registered and referred
to as Delta Sierra Juliet—follows. I have carefully studied the audiotape
and noted the many times where Valentich's voice inflections rise at the
end of his transmissions, as if he were asking a question. The young pilot
was clearly disoriented by 9:10 p.m. at the latest and probably earlier.
There are many pauses during his transmissions, which are indicated by
three ellipsis points.

9:06:14 Valentich: Melbourne, this is Delta Sierra Juliet. Is there any
known traffic below five thousand?

9:06:23 Robey: Delta Sierra Juliet—no known traffic.

9:06:26 V: Delta Sierra Juliet—I am—seems [to] be a large aircraft
below five thousand.

9:06:46 R: Delta Sierra Juliet—What type of aircraft is it?

9:06:50 V: Delta Sierra Juliet—I cannot affirm. It is four bright . . . it
seems to me like landing lights.

9:07:04 R: Delta Sierra Juliet.

9:07:32 V: Melbourne, this [is] Delta Sierra Juliet. The aircraft has
just passed over me at least a thousand feet above.

9:07:43 R: Delta Sierra Juliet—Roger—and it, it is a large
aircraft—confirm?

9:07:47 V: Er, unknown due to the speed it's travelling . . . Is there
any Air Force aircraft in the vicinity?

9:07:57 R: Delta Sierra Juliet. No known aircraft in the vicinity.

9:08:18 V: Melbourne—it's approaching now from due east—
towards me.

9:08:28 R: Delta Sierra Juliet.

9:08:49 V: Delta Sierra Juliet. It seems to me that he's playing some sort of game—he's flying over me two—three times at a time at speeds I could not identify.

9:09:02 R: Delta Sierra Juliet—Roger. What is your actual level?

9:09:06 V: My level is four and a half thousand, four five zero zero.

9:09:11 R: Delta Sierra Juliet . . . And confirm—you cannot identify the aircraft.

9:09:14 V: Affirmative.

9:09:18 R: Delta Sierra Juliet—Roger . . . Standby.

9:09:28 V: Melbourne—Delta Sierra Juliet. It's not an aircraft . . . It is . . .

9:09:46 R: Delta Sierra Juliet—Melbourne. Can you describe the . . . er, aircraft?

9:09:52 V: Delta Sierra Juliet . . . as it's flying past it's a long shape . . . [cannot] identify more than that. It has such speed . . . It is before me right now, Melbourne?

9:10:07 R: Delta Sierra Juliet—Roger. And how large would the, er, object be?

9:10:20 V: Delta Sierra Juliet—Melbourne. It seems like it's chasing me. What I'm doing right now is orbiting, and the thing is just orbiting on top of me also . . . It's got a green light and sort of metallic [like]. It's all shiny [on] the outside.

9:10:43 R: Delta Sierra Juliet.

9:10:48 V: Delta Sierra Juliet . . . it's just vanished.

9:10:57 R: Delta Sierra Juliet.

9:11:03 V: Melbourne, would you know what kind of aircraft I've got? It is [a type of] military aircraft?

9:11:08 R: Delta Sierra Juliet. Confirm the . . . er, aircraft just vanished.

9:11:14 V: Say again.

9:11:17 R: Delta Sierra Juliet. Is the aircraft still with you?

9:11:23 V: Delta Sierra Juliet . . . It's, ah . . . [now] approaching from the southwest.

9:11:37 R: Delta Sierra Juliet.

9:11:52 V: Delta Sierra Juliet—the engine is, is rough idling. I've got it set at twenty-three—twenty-four . . . and the thing is—coughing.

[Engine trouble is audible on the audio tape.]

9:12:04 R: Delta Sierra Juliet—Roger. What are your intentions?

9:12:09 V: My intentions are—ah . . . to go to King Island—ah, Melbourne, that strange aircraft is hovering on top of me again . . . it is hovering and it's not an aircraft.

9:12:22 R: Delta Sierra Juliet.

9:12:28 V: Delta Sierra Juliet—Melbourne . . .

[A pause for seventeen seconds during which a very strange, metallic-sounding pulsed noise is audible, with no discernable pattern in time or frequency.]

9:12:49 R: Delta Sierra Juliet, Melbourne.

End of transcript.

Valentich was never heard from again.

Valentich's description of "a green light and sort of metallic like, all shiny on the outside" is important. In the years following the event, a colleague obtained reports from twenty eyewitnesses in the region, describing an erratically moving *green* light in the sky at that same time of evening as Valentich's flight. Years later, I traveled to the resort town of Apollo Bay, Australia, and interviewed Ken Hansen who was forty-seven at the time of the incident in 1978, and his two nieces. Hansen was driving with the two girls when they noticed, in the sky above, the lights from a plane along with a large green light. The presence of that second light was so unusual that Hansen decided to pull over, stop, and get out of his automobile. He said that when he did so, he clearly saw a second large, greenish, circular light "like it was riding on top of the airplane." Its visual size, as he described it, was equivalent to that of a tennis ball held at arm's length, with a ratio between it and the plane of about four to one. Assuming this estimation is accurate, the UFO would be about forty-eight feet across. Its green color was similar to the navigation lights on an airplane. Hansen noticed that it kept a constant distance above and slightly behind the airplane's lights, while he watched for about fifteen to twenty seconds, until both lights disappeared from sight.

He told his wife that evening about the large green light, as well as his coworkers the next day, before he knew anything about what Valentich had reported. When we met, his nieces confirmed the details provided by their uncle. I was able to obtain much valuable information by going to the site with Hansen where he had pulled over in his car, because he reconstructed for us what he saw.

The story of Valentich's encounter with a UFO and subsequent disappearance was reported by the media throughout the world, garnering much attention. Despite the coordinated efforts of private pilots and the Australian government's search-and-rescue airplanes, no trace of him or his airplane has ever been found. There is sufficient evidence to suggest that he probably crashed into the sea between three and twelve miles offshore, but we will likely never know what happened. The nature of the large object with green lights that accompanied the airplane during its last minutes remains even more of a mystery.

About two months later, a remarkable aerial sighting was documented over New Zealand. Captain Bill Startup, a senior pilot working for Safe Air Ltd. with twenty-three years of experience and 14,000 hours of flying time, and his copilot Robert Guard, with 7,000 hours of flying time, were key witnesses. The Argosy freight plane they piloted was making a newspaper delivery between Wellington and Christchurch off the Kaikoura coast of South Island. Australian television reporter Quentin Fogarty from Channel O in Melbourne, Australia, his cameraman David Crockett, and sound operator Ngaire Crockett were also on board, because UAP had been witnessed by aircrews and picked up by radar about ten days earlier along the same route. Fogarty was making a television documentary about these earlier events, partly because of heightened interest in UFOs after the Valentich disappearance. He wanted background footage for his documentary, so he joined the newspaper delivery on December 30–31, 1978, for this purpose. He never expected to witness any strange phenomena himself.

But just after midnight on that flight, a series of light phenomena appeared, escorted the aircraft, and flew around it. Captain Startup and copilot Guard, who were well aware of the regular, very familiar lights along the coast, were the first to notice the strange lights ahead of them. For about thirty minutes, cameraman Crockett captured the luminous objects on 16 mm color movie film, while Fogarty commented on camera, as the events unfolded. At the same time, on-board systems and air traffic control in Wellington, New Zealand, tracked the objects on radar while they were viewed by Captain Startup and others aboard. The radar readings were reported to the pilots by air traffic controller Geoffrey

Causer, and witnessed on the scope also by radar maintenance technician Bryan Chalmers. Causer remained in constant communication with the pilots throughout the incident, and the entire dialogue was recorded on tape.

I have viewed the film of these unusual images—showing bright lights in and out of focus, some round, some suggestive of a disc shape—which has also been carefully analyzed by others. The lights disappeared and reappeared in totally new locations, sometimes several at a time. Their behavior cannot be explained by normal aerophysics.

At one point, witnesses in the plane observed lights flying in formation with the aircraft. They then heard from air traffic control that the phenomenon was so close to the plane that the radar scope could not separate the two. Causer registered only one signal on the radar screen, but it was twice as big as it had been before. "There's a strong target right in formation with you. Could be right or left. Your target has doubled in size," he reported. Chalmers also viewed the double-size target. It appeared as if two planes were flying at the same speed so close to each other that they were indistinguishable on radar. Such proximity could of course be a threat to aviation safety, but this aircraft suffered no ill effects.

These are unusual cases. Shorter events involving near misses are more common. On August 8, 1994, a commercial flight en route from Acapulco, Guerrero, to Mexico City, Mexico, almost collided with a UAP that darted out of a cloud straight toward the aircraft. Fortunately, the UAP maneuvered to avoid the collision. A Japanese Transocean Air Boeing 737 commercial airliner was on route from Okinawa Prefecture to Tokyo at cruising speed on November 11, 1998, when the first officer suddenly saw two white "strobe lights" in front of him. The two lights separated rapidly, and he made a dive to avoid a collision. In these two cases, neither object was detected by ground radar. In 2004, during the sunny afternoon approach of a commercial flight to Brazil's São Paulo airport, both crew members saw a self-luminous sphere ahead of them that remained at their altitude as they descended. The twin turboprop airplane had to bank sharply and dive to avoid a collision.

In America, the case of Captain Phil Schultz is exceptional—one that I personally investigated. I interviewed the captain extensively and received a six-page, handwritten Aerial Sighting Report from him.

Captain Schultz was piloting TWA flight 842 from San Francisco to John F. Kennedy Airport over Lake Michigan one bright clear summer day in 1981. Suddenly he saw a "large, round, silver metal object" with six jet-black "portholes" equally spaced around the circumference, which quickly "descended into the atmosphere from above." Captain Schultz and his copilot were so close to the object that it appeared as large as a grapefruit held at arm's length. Expecting a midair collision, they braced themselves for impact. The object then made a sharp, high-speed turn, avoiding the aircraft, and departed.

Schultz did not file a report with TWA, but instead worked diligently with me to accurately reconstruct the event in the cockpit of his aircraft. This allowed me to ascertain many important facts about the event. Its approach and departure speed was calculated to be about 1,000 mph, with a high G turn, as well. No shock wave or turbulence was felt at any time. The aircraft's autopilot remained coupled throughout the encounter, and no electromagnetic effects were noticed. The first officer saw the final two-thirds of the event, but the flight engineer did not see anything as a result of his position in the rear of the cabin. Chicago Center had no

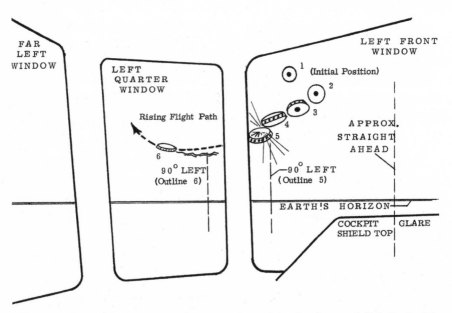

My sketch of the cockpit windows and apparent size, shape, location, and flight path of the UAP seen by Captain Schultz. R. Haines

other air traffic in the area, although their radar at the time had a range of about 150 miles.

With extensive experience as a U.S. Navy fighter pilot in the Korean War and afterward, Captain Shultz never accepted the reality of UFOs prior to this incident. This encounter instantly changed his belief. When I asked him what he thought the object was, he quickly replied, "There is no doubt in my mind. It was an extraterrestrial craft." He said as much in his handwritten report that he filled out for me, saying he believed the thing was a "spaceship."

Also in America, a very puzzling, low-altitude, in-flight apparent collision occurred on October 23, 2002, just northeast of Mobile, Alabama, according to a National Transportation Safety Board (NTSB) accident report. En route from Mobile to Montgomery, Thomas Preziose, fifty-four, with 4,000 total flight hours to his credit, was piloting alone, carrying about 420 pounds of paper records cargo. He took off for this flight at 7:40 p.m. The preliminary accident report stated that the Cessna 208B, a Cargomaster with the FAA registration number N76U—a high-wing, single-engine commercial airplane—"*collided in-flight with an unknown object* [italics mine] at 3,000 feet and descended uncontrolled into swampy water in the Big Bateau Bay in Spanish Fort, Alabama." The crash occurred about six minutes after take-off, at approximately 7:46 p.m. Interestingly, the NTSB saw fit to issue a later report that did not mention the collision with an unknown object.

Based on data from an automated surface-observing system 7.7 miles from the accident site, recorded at 6:53 p.m., there was a layer of scattered clouds at 700 feet and a more solid overcast beginning at 1,200 feet with clear air in between, and a visibility of five miles. The wind was 11 knots at 60 degrees. It may be significant to this fatal accident to note that a DC-10 passed about 1,000 feet above the Cessna after approaching him from about his eleven-o'clock position at 7:45—seconds before the crash—and would have produced wing-tip vortex turbulence. Afterward, the pilot uttered his final words before his death: "Night Ship 282, I needed to deviate, I needed to deviate, I needed to deviate, I needed—" (end of transmission at 7:45:57 p.m.).

If Preziose collided with a physical object, it was never located. Yet a strange red residue (referred to as "transfer marks") was found coating at least fourteen different areas of the downed airplane that were widely separated in location both inside and outside the aircraft. The engine block had been split, suggesting a very great force of impact. Unfortunately, radar data recording hardware was inoperative at the time of the accident, yet the NTSB did not request radar data from the Pensacola Naval Air Station, less than an hour away. The DC-10 that passed over the Cessna just before the crash was inspected upon landing, and no damage of any kind was found.

The final NTSB report indicated that the accident was caused by pilot disorientation. However, an independent investigation found numerous discrepancies with regard to both the FAA documentation and the investigation conducted by the NTSB.

Several samples of the red residue on the Cargomaster were analyzed using a Fourier Transform Infrared Spectroscopy device. One red sample was found to be most similar to reference material consisting of tere- and isophthalate polymer with the "possible presence of inorganic silicate compounds." Another sample of bare metal from the wing was found to be most similar to reference material consisting of "epoxy materials with some inorganic silicate fillers." While certain segments of metal from a U.S. Air Force unmanned aerial vehicle (UAV) were also subjected to the same analysis for comparison, little has been said of these findings except that their composition was "significantly different" from the red residue marks. The nearest Air Force base flying UAVs is Tyndall at Panama City, Florida, some 150 miles to the ESE.

If something struck this airplane, it certainly qualifies as a UAP until it is positively identified.

Considering the many kinds of UAP flight maneuvers that have been reported, it is clear that whatever the phenomenon is, it appears to be able to outperform high-performance aircraft in virtually every respect. This same conclusion was made in a recently unclassified report from the United Kingdom. In most of these pilot reports the aircraft appears to be the focus of "attention" of the phenomenon; this conjecture has been supported by many hundreds of high-quality foreign pilot reports as well. Hundreds of reports in my files suggest that the variety of phenomena are associated with a very high degree of intelligence and deliberate flight control.

The majority of pilot reports indicate that UAP tend to approach aircraft during darkness. At night, it is possible to see the readily discernable colors either within relatively small, localized regions (similar to individual light sources) and/or more diffusely over their entire surface. The appearance of the UAP's lighting patterns takes many different forms; they might be interpreted as some type of aircraft anticollision or navigation lights, even though intense blue lights, generally not permitted in America, are reported in some cases.

Most pilots understand that they will experience a wide range of visual phenomena in the atmosphere over the course of their flying career, but they do not expect that some will remain unexplained after considering all known natural phenomena and man-made objects. When this happens, each witness is left with a lingering uncertainty, a doubt about the core identity of what was seen, and must wrestle with a decision about whether or not to report the event.

Most likely, he or she will not do so. Pilots know how people are treated when discussing or reporting strange sightings, and they are not inclined to risk ridicule or job security. I call this the "law of diminishing reports"—a negative feedback effect that inhibits more and more people from saying anything about what they've seen. The long-term effect of this is that less and less reliable data becomes available for serious study, and the whole subject of UAP slides farther into the realm of myth and societal humor. Since this has been going on for many decades, airline administrators and government bureaucrats can validly claim that there is nothing to investigate or take seriously because pilots are not reporting anything. And scientists who rightly claim that they cannot study a phenomenon without having reliable data are justified for not becoming interested! Already rare "anomalous" phenomena seem to become even rarer, reinforcing the mistaken belief that these events don't occur in the first place.

Air traffic controllers are often aware of these unreported encounters with UAP, since they are normally the first to receive radio calls from the cockpit crew about the UAP, or pick up the targets on radar. But they, also, do not report many incidents. A controller at Los Angeles Air Route Traffic Control Center wrote, "In my six years at the Center, I have personally been part of three bizarre encounters, non-military and non-civilian. I'm just one of 15,000 controllers, too, so there have to be many more that go unreported . . . In a fourth incident I was present for (in the area but not at the actual sector), the controller told the supe about the encounter, and after both determined there was nothing on radar, they just kind of shook their heads and rubbed their chins, and that was that. This I believe is what typically happens. Nobody knows what to do, really."

Based on surveys and pilot interviews conducted by myself and associates at NARCAP, we estimate that only about 5 to 10 percent of pilot sightings of UAP are reported. Unless we implement policy changes, aircrew will continue to remain silent.

History is filled with accounts of previously ridiculed subjects that have turned out to be important to mankind, as a study of the history of science confirms. We must not simply overlook UAP because we are uncomfortable with the mere thought of them. Neither society's current prejudice toward UAP nor its abiding ignorance about them is likely to prevent their continued appearance, nor do such responses prove that they don't exist. These phenomena simply won't go away.

CHAPTER 6

Incursion at O'Hare Airport, 2006

On November 7, 2006, something unimaginable happened at Chicago's O'Hare Airport during the routine afternoon rush hour. For about five minutes, a disc-shaped object hovered quietly over the United Airlines terminal and then cut a sharp hole in the cloud bank above while zooming off. Hardly anyone heard about it until the story broke on the front page of the *Chicago Tribune* on January 1, 2007, almost two months later, which precipitated a flurry of national coverage on CNN, MSNBC, and other networks. With over a million hits, the *Tribune*'s story quickly achieved the status of being the most-read piece in the entire history of the newspaper's website, but then faded from the media radar screen. No official assessment was ever provided to a fascinated but alarmed frequent-flying public or to the employees of United who were directly involved.

It was an ordinary, overcast day, with visibility of about 4 miles and winds at 4 knots. Between 4:00 and 4:30 p.m. on that afternoon, pilots, managers, and mechanics from United Airlines looked up from their ground positions at the terminal and saw the strange object hovering just under a cloud bank, which began at 1,900 feet above the ground. According to these witnesses, the metallic-looking disc was about the size of a quarter or half dollar held at arm's length. Based on the collection of eyewitness testimony, the UFO is estimated to have ranged in size from about 22 to 88 feet in diameter, and was suspended at approximately 1,500 feet over Gate C17 at the United terminal.

A pilot announced the sighting over in-bound ground radio for all grounded planes; a United taxi mechanic moving a Boeing 777 heard radio chatter about the flying disc and looked up; pilots waiting to take off opened the front windows to lean out and see the object for themselves. There was a buzz at United Airlines. One management employee received a radio call about the hovering object, and ran outside to view it for himself. He then called the United operations center, made sure the FAA was contacted, and drove out on the concourse to speak directly with witnesses there.

Reports show the event lasted from about five to fifteen minutes. Then, with many eyes now fixated on it, the suspended disc suddenly shot up at an incredible speed and was gone in less than a second, leaving a crisp, cookie-cutter-like hole in the dense clouds. The opening was approximately the same size as the object, and those directly underneath it could see blue sky visible on the other side. After a few minutes the break in the cloud bank closed up as the clouds drifted back together. "This was extremely unusual, according to the witnesses," *Chicago Tribune* transportation reporter Jon Hilkevitch told television news after interviewing the United witnesses for his story. "Airplanes just don't react like this. They slice through clouds."

This was definitely not an airplane, the observers said, and many seemed shaken by what they had seen. Some were awed; others afraid. "The witness credibility is beyond question, and safety was a big concern," Hilkevitch said during a phone conversation. He noted that all observers independently described the same thing: a hovering disc making no noise as it shot up and left a clear hole in the clouds. "The only discrepancies were their size estimations and that some said it was rotating or spinning," he told me.

Sadly, every one of these highly credible aviation witnesses to the O'Hare UFO—and there were many—has chosen to remain anonymous, due to fears for job security. One United employee told me he could otherwise be perceived as "betraying" his company. Witnesses do not want to be "caught talking to the media since the airline had officially claimed that nothing happened," he wrote in an e-mail. These witnesses to something that's not supposed to exist—something laughed at by their colleagues—were left alone with their unsettling observations. "I realize this is a controversial position, but with my extensive knowledge of modern aviation technologies, I know this UFO probably wasn't created on this planet," one told me a few months afterward.

The FAA and United Airlines at first denied having any information about the incident, but both had to acknowledge the sighting when a tape of a United supervisor's call to the air traffic control tower was released by the FAA.

I have listened to those tapes.

"Hey, did you see a flying disc out by C17?" asked the supervisor, giving her name as Sue. Laughter is audible from tower operator Dave and a second person nearby. "That's what a pilot in the ramp area at C17 told us," she continues. "They saw some flying disc above them. But we can't see above us." The laughter continues nervously, and Dave replies, "Hey, you guys been celebrating the holidays or anything, or what? You're celebrating Christmas today? I haven't seen anything, Sue, and if I did I wouldn't admit to it. No, I have not seen any flying disc at gate C17."

About fifteen minutes later, Sue calls back again, this time reaching operator Dwight. The conversation is as follows:

Sue: "This is Sue from United." (laughter)
Tower: "Yes." (serious tone)
(12 second pause)
S: "There *was* a disc out there flying around."
T: "There was a what?"
S: "A disc."
T: "A disc?"
S: "Yeah."
T: "Can you hang on one second?"
S: "Sure."
(pause, 33 seconds)
T: "Okay, I'm sorry, what can I do for you?"
S: "I'm sorry, there was, I told Dave, there was a disc flying outside above Charley 17 and he thought I was pretty much high. But, um, I'm not high and I'm not drinking."
T: "Yeah."
S: "So, someone got a picture of it. So if you guys see it out there—"
T: "A disc, like a Frisbee?"
S: "Like a UFO type thing."
T: "Yeah, okay."
S: "He got a picture of it." (laughs)
T: "How, how, how high above Charley 17?"

S: "Well, it was above our tower. So . . ."
T: "Yeah."
S: "So, if you happen to see anything . . . " (she continues to laugh)
T: "You know, I'll keep a peeled eye for that."
S: "Okay."

Unfortunately, the photograph Sue referenced has never been located. Also, due to the way the towers were constructed, the operators were not able to see the UFO; its location in the sky was not within their visual field through the glass window because of the roof overhang, so it hovered in what amounted to the tower's blind spot. Planes full of passengers were landing and taking off while the "UFO type thing" sat poised in the sky overhead, and no one knew what this thing was, why it was there, or what it might do next. This taped exchange, which includes giggling, Sue's need to proclaim she wasn't "high," and Dave's admission that he wouldn't admit it even if he had seen the disc, is a glaring commentary on the UFO taboo that infects aviation personnel even in the midst of an ongoing, possibly dangerous incident being reported by trained observers of aircraft.

Dave might have reacted differently if the flying disc had been picked up on radar, but it wasn't. Perhaps the object had some kind of stealth capability, but at the same time we know that airport radars are not configured to register stationary objects such as this, or, at the other extreme, extraordinarily high-speed motion, because such behavior is outside the norm. The O'Hare incident is not the only example of this. Unidentified objects are often not detected on radar, even when physically present and seen by multiple witnesses, and obviously this doesn't mean they aren't there. In many other cases, radar tracks *are* captured, providing valuable data on the object's movements. What determines this variability in detection is unknown.

Fortunately, a team of experts from Dr. Richard Haines's group NAR-CAP spent five months rigorously investigating the incident and its safety implications, and analyzing all possible explanations for the sighting. Their 154-page report was co-authored by Haines; meteorologist William Pucket, formerly with the Environmental Protection Agency; aerospace engineer Laurence Lemke, also previously with NASA on advanced space mission projects; Donald Ledger, a Canadian pilot and aviation professional; and five other specialists. They concluded that the O'Hare disc was a solid physical object behaving in ways that could not be explained in conventional terms. It had penetrated Class B restricted airspace over a major airport without utilizing a transponder.

The NARCAP study stated:

> This incident is typical of many others before it in that an unknown phenomenon was able to avoid radar contact and, thus, official recognition and effective response. When combined with the deeply entrenched bias pilots have against reporting these sightings, the FAA seemingly had justifiable grounds for ignoring this particular UAP as non-existent.

And indeed the FAA tried hard to ignore the incident despite its safety implications, but pressure from the *Chicago Tribune* and others forced a response. Initially an FAA spokesperson attempted to explain the incident as airport lights reflecting off the bottom of the cloud ceiling. However, the event occurred in daylight and the airport lights hadn't been turned on yet! In a second try, a different spokesperson wrote the whole thing off as a "weather phenomenon." Obviously, these United pilots and airport employees know how to recognize airport lights on clouds and unusual weather conditions, though it was a normal overcast day. They would not have described a flying disc, each providing the same independent description from different vantage points, if some strange weather was unfolding, and to suggest otherwise is an insult to those doing their duty by reporting the incursion.

Transportation expert Hilkevitch, who routinely covers the much less exciting, mundane events that occur on a regular basis at O'Hare Airport, was mystified by the FAA disinterest in the incident. "If this had been a plane, it would have been investigated," he told me. "The FAA treats the smallest safety issue as very important. It will investigate a coffeepot getting loose in the galley and falling while a plane is landing." Brian E. Smith, a former manager within NASA's Aviation Safety Program, told me that "managers should want to hear about such vehicle operations before they become accidents or disasters." He said the safety implications of *anything* operating outside the authority of air traffic control at a major airport are obvious, no matter what type of vehicle it is.

The NARCAP experts concurred:

> Anytime an airborne object can hover for several minutes over
> a busy airport but not be registered on radar or seen visually
> from the control tower, it constitutes a potential threat to flight
> safety. The identity of the UAP remains unknown. An official
> government inquiry should be carried out to evaluate whether
> or not current sensing technologies are adequate to insure
> against a future incident such as this.

So, what exactly was going on here?

I decided to call FAA spokesperson Tony Molinaro and ask him for
more details about the bizarre "weather" that he said United Airlines
pilots mistook for a physical object—weather so freakish that it was able
to cut a round, sharply defined hole though a thick cloud bank in a split
second. Such a phenomenon would certainly be worthy of study by scientists in the age of climate change, and is actually even more of a novelty
than hovering or speeding discs, which have made the news since the
1940s.

"In the absence of any kind of factual evidence, there is nothing
more we can do," Molinaro said in a phone interview, in response to my
asking why the FAA chose not to investigate this. But was there factual
evidence for his newly discovered weather phenomenon? Weather is the
best guess, he said, and then pointed to a specific natural phenomenon
that isn't really weather: a "hole-punch cloud," as it is colloquially called.
After all, he stated, such a cloud hole is in "a perfect circular shape like
a round disc" and has "vapor going up into it." In other words, witnesses
mistook the cloud hole for a disc (even though the disc was seen for
many minutes before the hole was created), and the *ascension* of vapor,
somehow moving up in defiance of gravity, was what witnesses believed to
be the disc shooting upward through the clouds.

Doesn't this sound ridiculous, if you stop and think about it? It's
the kind of response that has typically been provided for decades when
officials are pressured to say *something*. And even if Molinaro hedged his
explanation by qualifying it as a "guess," this kind of subtle understatement is quickly lost to the mass media and the general public.

And was his guess at all reasonable? I contacted weather experts and scientists specializing in cloud physics, something the FAA would have been wise to have done before issuing its explanation. No, this could not possibly be what witnesses saw, I learned.

Hole-punch clouds are formed when ice crystals from a higher cloud deck fall onto a lower one. The hole is formed by ice crystals falling *downward,* not upward as Molinaro postulated. Super-cooled water droplets in the lower cloud adhere to the crystals, enlarging them and leaving a space around them in the cloud. The crystal mass accumulates weight and then falls farther, below the second cloud, evaporating when it hits warmer air.

The key factor is that this process can only happen at *below* freezing temperatures. The temperature at 1,900 feet above O'Hare Airport the day of the sighting was 53 degrees F, according to the National Weather Service. The climatologists and other weather experts I spoke to all stated that temperatures must be below freezing for a hole-punch cloud to explain the sighting.

And they told me that a hole in a cloud can be formed by only one other means: evaporation by heat. And this just happens to fit the witnesses' explanation of what they saw: a high-energy, round object very likely to be emitting some form of intense radiation or heat while cutting through the cloud bank. Thus, isn't evaporation by heat the most logical explanation, the "best guess," for what happened?

The NARCAP team also recognized the folly of Molinaro's explanation:

> We postulate that the instantaneous nature of the hole formation, the circular shape, and its sharp edges all point to the direct emission of, for example, electromagnetic radiation from the surface of the oblate spheroid as the proximate cause of the hole in the clouds. We cannot identify the object or phenomenon lying inside the oblate spheroid surface, but two conclusions seem inescapable: (1) the object or phenomenon observed would have to have been something objectively and externally real to create the hole effect; and, (2) the hole phenomenon associated with this object cannot be explained by either conventional weather phenomena or conventional aerospace craft, whether acknowledged or unacknowledged.

Unfortunately, our government is not willing to issue a sensible statement about what actually happened, giving due respect to witness reports, and instead refuses to investigate. Once again, a curious general public is left out in the cold, frustrated, alarmed, and perplexed by their government's silence. In keeping with the historical pattern, the FAA's explanation of a hole-punch is factually ludicrous, since the temperatures at O'Hare were too warm for it to have even been a physical possibility.

Nonetheless, once the FAA explanation is tossed out and printed by the media, no matter how far-fetched, it provides a handy way out for those inclined to dismiss any and all UFO sightings, those committed to believing they don't exist. Most people will never know that temperatures at O'Hare render the FAA explanation impossible (this information was not put forward until months after the fact) and will be swayed by what the authorities tell them. The case from then on is tainted by that seed of doubt, which becomes part of the record. Those who *do* know the facts about the O'Hare incident continue to mistrust our government, which has demonstrated, once again, that it will avoid dealing with UFO incidents at all costs.

This recent event clearly illustrates the fundamental tenets about the UFO problem that I spelled out in the introduction: UFOs are real, physical objects; they remain unexplained; they can be an aviation safety hazard; our government routinely ignores them, disrespecting expert witnesses and issuing false explanations; the extraterrestrial hypothesis cannot be ruled out when no man-made or natural explanation applies; and an immediate investigation is required.

Why is our government uninterested in a strange, highly technological object hovering over a major airport, as reported by competent airline personnel? What about passenger safety? Or national security after 9/11? Or just plain scientific curiosity about an unexplained phenomenon? Official distate for dealing with the UFO phenomenon is entrenched to the point of being not only counterproductive, but possibly dangerous.

CHAPTER 7

Gigantic UFOs over the English Channel, 2007
by Captain Ray Bowyer

Five months after the O'Hare incident, on April 23, 2007, another sighting occurred involving pilots and aviation personnel, this time over the English Channel off the French coast of Normandy. Commercial airline pilot Ray Bowyer did not hesitate to report his sighting of two massive UFOs, witnessed by him and his passengers, even though it had no direct impact on flight safety. Following regulations, he submitted his report to the Civil Aviation Authority (CAA), Britain's aviation regulatory body responsible for air safety, the equivalent of our FAA. This time the objects were tracked on radar, and the sighting made news around the world, without delay.

Captain Bowyer says that there were no negative effects as a result of his speaking out about the incident when he was approached by the BBC. His airline offered every support he needed, and the local air traffic control released recorded information to journalists and researchers who asked about the case. "I did not feel that I was in any danger of being ridiculed, because all I did was report what actually happened, as was my duty," he stated.

Especially after learning about the O'Hare Airport case, which occurred only months before his sighting, Bowyer noted the differences between the British and U.S. reporting systems, and also between the official attitudes within the two countries. The fact that crews and ground personnel were pressured by their company not to discuss the incident, and that the FAA did not investigate, surprised him. "I would have been shocked if I was told that the CAA would not be investigating, or if the CAA told me that what I had seen was something entirely different," he commented in response to the FAA's claim that witnesses were actually observing weather. "But it seems that pilots in America are used to this kind of thing, as far as I can tell."

I first met Captain Bowyer at our Washington, D.C., press conference six months after his sighting, when I also met General De Brouwer. He attended for a few days with the full cooperation of his airline, Aurigny Air Services, which flies between the Channel Islands and both France and the UK. I found Bowyer to be a remarkably frank, down-to-earth, utterly incorruptible British everyman; in other words, a naturally honest man, blessed also with a great sense of humor. His account that follows, though at times alarming, gives expression to these personal qualities, and stands in interesting contrast to the more formal and restrained writing styles of our military contributors.

There has always been a strong connection to flying in my family, and even though I initially trained as a production and research engineer, I always had a hankering to get airborne. So in 1985 I began to fly, and four years later I qualified as a commercial pilot. Since then, I have worked for many airlines in Britain, Europe, and the Middle East.

I spent ten years, beginning in 1999, with Aurigny Air Services, based in the Channel Islands, which lie between southern Great Britain and northern France. Aurigny flies between the three largest islands—Alderney, Jersey, and Guernsey—and western France and England. I have completed some 5,000 hours and 8,000 landings for Aurigny in Britten-Norman Trislander aircraft. Although very basic and rather noisy, these eighteen-seat, three-engined aeroplanes are strong and ideal for short-sector work into short runways such as those at Alderney, the most northerly and smallest of the islands serviced by the airline. The flight deck area of the Trislander is not separated from passengers—we all sit in essentially one open cabin. While piloting the aircraft, I can literally turn around and talk to the passenger behind me.

On April 23, 2007, my passengers and I witnessed multiple, as yet unidentified objects over these islands while crossing the English Channel. They were very, very large. The objects were picked up on radar in two locations, and one was witnessed by another pilot from a totally different vantage point.

At 4,000 feet on that afternoon, the visibility was very good—at least

100 miles all around—with a low-level haze layer underneath us up to 2,000 feet. We were on route from Southampton, England, to Alderney, which takes about forty minutes, cruising at 150 mph.

At first I saw one object that seemed close because of its apparent size, and I considered it to be only five or six miles distant. However, as time passed with the object remaining in view, even though I had flown twenty miles closer to it, it still appeared to be a good distance away.

When I first saw it, I thought, based on past experience, that this brilliant yellow light was a reflection of the sun from a commercial greenhouse in Guernsey, famous for its production of tomatoes. But in this case the relative motion of the aircraft in combination with the critical angle between the ground and the sun meant that such a reflection could not occur. Furthermore, there was no direct sunlight from above as there was a layer of cloud at 10,000 feet covering the whole area. With this in mind, I reached for my binoculars while flying on autopilot, and viewing it magnified ten times, found that this light-emitting object had a definite shape: that of a thin cigar, or a CD viewed on edge with a slight incline. It was sharply defined, and pointed on both ends. The aspect ratio was approximately 15:1 and I could clearly see a dark band two-thirds of the way along from left to right while viewing it through the binoculars.

As I drew nearer to the object, a second identical shape appeared beyond the first. Both objects were of a flattened disk shape with the same dark area to the right side. They were brilliant yellow with light emanating from them. I passed the information to Jersey air traffic control (ATC) and they initially said they had no contact. I pressed the point over the next few miles and the controller at Jersey, Paul Kelly, then said he had primary contacts south of Alderney. So here we were on a bright afternoon in May with two objects ahead getting closer and larger with no explanation as to what they were! I found myself astounded, but curious.

At this point, the passengers began to notice the unusual things and to ask about them. I decided not to make any announcement over the intercom so as not to alarm anyone, but it was obvious that some were getting concerned. By now the two identical objects were easily visible without binoculars, the second one behind the closer one, with exactly the same characteristics albeit farther away.

ATC then informed me that there were two reflections from primary radar, both to the southwest of Alderney. This was beyond my destination, for which I was glad as the objects were becoming uncomfortably close. Their brilliance is difficult to describe, but I was able to look at this fantastic light without discomfort. They both seemed to be stationary, but the radar traces later proved otherwise: they were actually moving away from each other at about 6 knots, one to the north, from the northern tip of Guernsey towards Casquettes lighthouse, the other moving south along the northwest coast of Guernsey.

Due to the haze layer it is unlikely that the objects were visible from the ground; however, after the event BBC radio received one uncorroborated report that one had been seen by a tourist staying at a local hotel in Sark, close to the Casquettes lighthouse.

Approaching the point to begin descent, twenty miles NNE of Alderney, I maintained an altitude of 4,000 feet to remain in good view of the objects. If they started to move off, I wanted to be able to take action to avoid them if at all possible.

Due to my close proximity, the dark area on the right of the nearest one now took on a different appearance at the boundary between the brilliant yellow and the dark vertical band. There appeared to be a pulsating boundary layer between the two differences in color, some sort of interface with sparkling blues, greens, and other hues strobing up and down about once every second or so. This was fascinating, but I was now well beyond our descent point and to be frank I was not too displeased to be landing.

My feelings at this time were mixed. The safety of the passengers is paramount and that always comes first, so to land was the priority. However, I was really intrigued with whatever was ahead of me, even though I was healthily trepidatious as well. If the aircraft had been empty, I would have gone a lot closer, perhaps overflown the nearest object to gather further information and satisfy my curiosity. However, I would never knowingly put passengers at risk. My last sight of the objects was whilst passing through 2,000 feet in the descent through the haze layer.

Throughout the whole encounter, which lasted fifteen minutes, there had been no interference with any of the aircraft systems or instruments, and radio communications were likewise unaffected.

Upon landing I asked if any of the passengers had seen anything unusual, without wishing to lead them, and told them if they had and should want to report it, to leave their name and number at the check-in desk. Passengers Kate and John Russell, sitting three rows behind me, went public with their sightings and their story is well documented. At least four other passengers saw the objects and the gentleman sitting behind me even borrowed the binoculars for a closer look.

I walked to our operations department to make an official report, as required by law, informing the powers that be that unidentified aircraft had been seen within controlled airspace where they certainly shouldn't have been. I drew a brief sketch and this was sent to Jersey ATC and onward to both the Ministry of Defence and the Civil Aviation Authority in London. With that done, it was time to grab a quick cup of tea and return to Southampton with another load of passengers.

I was somewhat concerned at the thought of departing to the west toward where I had last seen the closest object, and although nothing was visible ahead whilst I lined up on the runway, I was aware that I had lost contact with the pair only due to the haze layer. Thankfully, after passing above 2,000 feet, there was nothing to be seen.

It was then, on this trip back to Southampton, that I had time to take stock of how big the two objects actually were. While in Alderney, I had received confirmation of the radar traces from the controller who had reviewed the data. I was able to determine that I was approximately fifty-five miles away from the first object, not the ten miles or less that

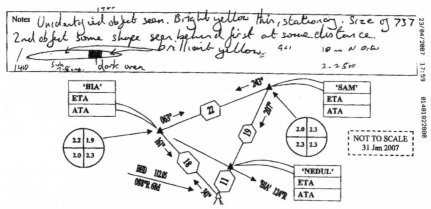

The section of my report, filed immediately after the incident, which included my drawing of one object. It was sent to the CAA and the MoD before I realized how large the objects actually were. R. Bowyer

I had originally thought. Flying around Europe at night, one gets to know the size of towns and cities relative to specific ranges, putting a scale on places of known size, along with a known oblique angle from a distant viewpoint. I was able to apply this same reference to the unidentified objects, presuming that they were flattened discs; they of course appeared long and thin from my viewpoint from the side. Seeing a reasonably large town from fifty-five miles would have been comparable to the size of this object. It was at this point that its massive size became clear, and I estimated it to be up to a mile long.

On my subsequent return to Alderney from Southampton, I telephoned Jersey ATC and spoke to Paul Kelly, the duty controller who was in communication with me during the sighting. He informed me that a pilot from a second aircraft had described a sighting as "matching the description" of what I had seen. This was a great relief to me, as it confirmed that I alone was not bonkers!

Indeed, Captain Patrick Patterson, the pilot of a Blue Islands Jetstream aircraft inbound to Jersey from the Isle of Man, had witnessed the same thing as me, from twenty miles south above the tiny island of Sark. Some months later, I met with Captain Patterson and we exchanged views as to what we had seen. Although his sighting was only for one minute or so, his description was proof to me that we had seen the same thing, even though he saw only a single object, the second being in his six-o'clock position and therefore out of view.

The decluttered radar trace recorded at the time clearly shows two slow-moving objects appearing simultaneously and disappearing off the trace simultaneously. The traces begin and end at exactly the same time, not a minute apart or even ten seconds. The northernmost of the two objects ends up in its final moments transiting overhead of the Casquettes lighthouse. The radar also shows the Blue Island aircraft top left to bottom right and my aircraft top right to centre.

A lengthy report by a team of independent researchers (with which I partly disagree on some content) overall offers no evidence to explain the sighting, which confirms to me that two tangible objects did appear over the Channel Islands that day. This study goes into extraordinary detail and runs over 175 pages, with references to the weather, temperature inversions, military activity, surface shipping movements, and many other avenues of investigation. I do, however, have a significant point at which I have to disagree with the team, which is their dismissal of the radar traces as being returns probably from a cargo boat.

Why would the two traces start and stop in midocean, at exactly the same moment, when they should be seen leaving or returning to port? And the northern object ends up in its final moments transiting over the Casquettes lighthouse, the scene of many shipwrecks including the SS *Stella* in the late nineteenth century with great loss of life. With tides running to 8 knots in this area, this surely would be a most inappropriate, indeed foolhardy, place to navigate a cargo vessel!

Regardless of the controversy, and even though many sightings by pilots do not have multiple witnesses or radar tracking, I would still urge all aircrew to report whatever they see as soon as possible, and to stand up and be counted.

Air law stipulates quite clearly that if an operating crew of an aircraft sees another aircraft in a place that it should not be, then at the earliest opportunity the whole scenario is to be reported to the relevant authority. In my case the British Civil Aviation Authority knew within twenty minutes of the sighting what was seen, as described in the flight log faxed directly to the relevant CAA office. The military was informed by Jersey air traffic control at the same time. This is not an option; it is an obligation that crews react in this manner.

Ever since I saw the two UFOs and reported them openly, I have been asked the one question that everyone wants answered on this subject: "So what did you see, then? What do you think it was?" In truth I still don't have an answer that satisfies me.

There are a number of what-if's to be considered. For example, what if it had been nighttime? Or if there had been no cloud layer between the ground and the massive objects? Their sheer brilliance would surely have been seen from hundreds of miles away by people on the ground and all overflying pilots. The surface of the sea and land would have been lit up as if by two mini-suns.

Also, four days later, there was an earthquake off the Kent coast some 200 miles away. Could they have been earthquake lights, a rare phenomenon coincident with earth tremors? Unlikely. These are not seen over water, since they are discharged directly from a fault line. And could one of them manifest as a stationary, brilliant, sharply defined object, with an exact duplicate some distance away? Highly doubtful.

Or, was the brilliance of the objects just an aside, since perhaps they were part of some secret experiment? I would be interested to know if an overflying military or government satellite had picked up this extraordinary power source or brilliant light, which would seem likely. In any case, the Ministry of Defence stated in writing that this was not a military exercise or anything belonging to them.

My conclusion for all that ask the question is simple: I believe that there were two solid craft working in unison that day, shown by the fact that their sortie was linked together in both time and space. What they were, I cannot answer. What they were doing, once again I cannot answer. What I will say is that for machines so huge that they are visible from 50-plus miles hence, from two independent sources and with radar evidence to support their provenance, I can only conclude that they were not from around here, and in that I mean they were not, can not, have been manufactured on Earth.

So, what's next? Well, this case, like so many others, was closed before it even started, as far as the authorities are concerned. The British and French military showed their now customary "Not too worried, really" colors, since their respective airspaces were not under direct threat. I interpret this to mean "We see it, but there's not a thing we can do to stop it or make it go away."

I believe that what we witnessed that day, along with what many other pilots witness around the world on a regular basis, *is* known to the relevant authorities as something not originating on this planet, and this has been known for a very long time.

But what if the people of the world were informed of this? It could result in recrimination against government, religion, and authority, possibly large-scale civil unrest culminating in a new world order which might or might not be beneficial to the planet, or a myriad of other complicated and unpredictable scenarios. The authorities may do well to consider keeping the lid on Pandora's box at this time.

On the other hand, I believe the time is coming when they will no longer be able to keep sweeping this issue under the mat. With improved technology available and with more sightings being faithfully recorded every day, the time is surely not far away. Soon they will have to confront the people with what is known. Depending on what they know, or what we might be able to learn once they do this, I suspect this might turn out to be the time when the human race will grow up. Forced to confront their own smallness in relation to Earth's place in the universe, humans may at last face up to a future as a tiny fish in a big sea.

This whole episode has exposed a new world to me that I didn't know existed. I've come to know a very unusual group of people who are fascinated with the subject of UFOs—an eclectic bunch of dedicated believers and dreamers, writers, skeptics, filmmakers, witnesses, psychotherapists, former military officials, and all hues in between. Some of the people that I have met firmly believe in extraterrestrial intelligence; others insist on refuting any idea of a greater intelligence than the human mind. Either way, the beliefs are firmly held and vocally expressed in all forms of media. And a complete industry has emerged to service the hunger for knowledge surrounding this subject.

As for myself, life has returned pretty much to normal. I still give the occasional newspaper, TV, or radio interview, but here at home in Guernsey, the incident is mainly forgotten. People have other things on their minds now, and concern about something otherworldly when the mortgage payment is due falls firmly into second or third place. Regardless, the day may be coming when the whole human race will have to face the frightening reality that we are coexistent with others in this universe. In my view, we may be well advised to get along with this now, because, frankly, we have very little choice.

CHAPTER 8

UFOs as Air Force Targets

C ommercial passenger jets operate quite independently from military aircraft, and obviously, as described by Richard Haines, have limited options when it comes to responding to a nearby UFO. Also, at least in America, the stigma against reporting such events is high among civilian pilots, who face the possibility that, if reported, the story might be leaked to the media, compounding the derision even further. Not one witness would go on the record regarding the 2006 O'Hare case, despite the numbers who validated the incident and despite the legitimate concern about aviation safety expressed by many of them. But what happens when pilots in military jets, fully armed, encounter UFOs? Or, if electromagnetic radiation from UFOs disables sensitive equipment at military bases, as it can do in the cockpit of an airplane, does this become an issue of national security? These considerations go one step further than that of aviation safety problems caused by accidental proximity to UFOs. When is it appropriate for military jets to take aggressive action, if ever?

As contrasted with commercial aviation, the military operates within a more self-contained, less public arena. Unlike commercial pilots, who are committed to assuring the comfort and safety of often hundreds of passengers as well as protecting their personal reputations and the reputations of their parent companies, Air Force officers have a very different set of priorities. Instead, these pilots are oriented toward protecting their homelands from attack and maintaining readiness for an unanticipated invasion or terrorist assault. Military fliers are prepared to defend themselves if necessary; their jet aircraft are loaded not with passengers but with lethal weaponry that can be used either to attack or to defend.

Military pilots and their air-traffic controllers are trained to obey orders and not ask too many questions, and the system is well practiced in the arts of both reporting sensitive information and maintaining its confidentiality. Within the armed services, pilots are more likely to file reports as a matter of duty, free of the risks that commercial pilots face, because they know that such information will likely be restricted. When Air Force pilots are faced with a UFO, there are often other witnesses from a second aircraft or a base below, and information can be quickly relayed up the chain of command. These officers know that other aircraft can readily be scrambled as support in response to any unusual engagement. And they can defend themselves instantly if necessary.

Knowing this, one naturally wonders: Have military pilots ever shot at UFOs? The shocking answer is yes.

In November 2007, I was fortunate to meet and spend a few days with two pilots who have both engaged in lengthy "dogfights" with targeted UFOs. Retired Iranian general Parviz Jafari was a major in the Iranian Air Force in 1976 when he was ordered by the Air Force Command to man his Phantom F-4 II jet and approach a luminous UFO observed over Tehran. Several times during a wild cat-and-mouse chase, he and his backseat navigator attempted to launch a Sidewinder missile at additional smaller objects heading their way, but at the moment of fire their equipment shut down, returning to normal only when their jet moved away. The main object had been pursued by a second Air Force jet, was recorded on cockpit radar, and was observed from the ground by a general and experienced air navigation crews.

A second, similar event occurred four years later, in 1980, over an air base in Peru, when then Lieutenant Oscar Santa María Huertas was ordered to intercept what was at first believed to be an aerial spying device. He fired at the balloonlike object and barraged it with machine-gun shells, but they had no effect. He quickly realized this was something unknown, a UFO. Three different times he locked on to the object to fire when it was stationary, but each time, at the last instant, it shot straight upward. This UFO was witnessed in broad daylight by over a thousand soldiers and staff at the La Joya military base.

General Jafari and Comandante Santa María met for the first time at our 2007 press conference in Washington, D.C., also attended by General De Brouwer, Captain Ray Bowyer, and a number of other contributors to this book. This was an opportunity to present statements publicly, but it was also a unique opportunity for these men to converse over the course of a few days, forming the basis of an international network.

As the co-organizer and media contact for the event, and host for our panelists, I was privy to many private discussions over morning coffee and some that lasted late into the night. I will never forget the evening two days before the press conference when General Jafari and Comandante Santa María shook hands and sat down together for the first time. They had just arrived at the Washington Hotel after long journeys from very distant parts of the globe. These two unassuming gentlemen joined a small group of us at the hotel's rooftop restaurant, weary but relieved to be among friends and excited about the momentous press event that lay ahead. General Jafari, sitting to my right, was affable and animated, and soon was responding to a host of questions from those at our table about the 1976 incident. Neither Jafari nor Santa María knew much about the other's experience, and the conversation that followed was unplanned and spontaneous, with no tape recorders or cameras present to curtail its intimacy.

Comandante Santa María did not speak English, but shortly after Jafari began his account, a Spanish-speaking couple at the next table confessed that they couldn't help eavesdropping, and one of them offered to translate for him. Following Jafari, he told his story, prompted by questions from those around him. Both men, each a witness to one of the most unusual events in Air Force history, discovered through the ensuing exchange how similar their experiences were. Each could identify with the fear and awe expressed by the other in the retelling of his story. As Air Force pilots on two different continents, they both had suddenly found themselves face-to-face with something utterly impossible, yet powerfully real. It was one of the most remarkable, and chilling, few hours I've spent since beginning this journey ten years ago, and I felt privileged to bear witness to it. Both retired military men were humble, understated, and direct, as well as entirely believable. Jafari described one speeding object coming after his Phantom F-4 jet as he prepared to return to the base. Someone at our table asked him how he felt. "At that moment," he replied in his imperfect but colorful English, "I doubled my scare." Santa María made a drawing of his UFO on a sugar pack served with our coffee, which I saved as a memento.

And why had the two pilots felt compelled to fire at these UFOs?

General Jafari explained that he was acting in self-defense. He initially had no intention of taking any such action, because the Iranian general who ordered him and his navigator aloft was simply interested in getting a better look at the brilliant starlike object, to try to determine its identity. But Jafari soon found himself confronted with actions highly unexpected and threatening to his aircraft. Santa María's circumstances were different. At the outset, he was told the purpose of his mission was to destroy the "espionage device" above his air base, since it had failed to respond to normal communications. Neither pilot realized how futile his actions would be when attempting to fire at a UFO.

In retrospect, there will always be a question as to whether actual aggression was displayed by the UFOs, and we have no idea as to their intention or purpose, or even whether these concepts apply. However, such incidents, although rare, do raise serious national security questions. As it stands now, there appears to be uniform agreement at the highest military levels that UFOs are not belligerent. Even when provoked by human aggression, they do not retaliate—and we have to assume they have every capability of doing so. As General Denis Letty of France assured readers in the COMETA Report, although "intimidation maneuvers have been confirmed," UFOs have demonstrated no hostile acts to date.

Perhaps the real national security problem lies with impulsive, even if understandable, attempts by military pilots to defend themselves against what they soon discover are phenomena of vastly superior technology with unknown agendas—a truly frightening prospect. But even if pilots feel that self-defense is warranted, such actions could have disastrous consequences if they were ever successful in damaging their target, or if the object *did* respond aggressively after an attempt to destroy it. The risks in engaging militarily with something this powerful, and completely unknown, are self-evident. No one can predict the behavior of something we don't understand. Being in attack mode also diminishes the possibility of establishing communication with the UFO, if that were possible, or of simply learning more about it through cautious observation at close range. The accounts of Jafari and Santa María give the inside stories of what two Air Force pilots experienced when attempting to shoot down a UFO. They had received no training or any preparation for dealing with such an unanticipated eventuality.

CHAPTER 9

Dogfight over Tehran

by General Parviz Jafari (Ret.), Iranian Air Force

At about 11:00 p.m. on the evening of September 18, 1976, citizens were frightened by the circling of an unknown object over Tehran at a low altitude. It looked similar to a star, but bigger and brighter. Some called the air traffic control tower at Mehrebad Airport, where Houssain Pirouzi was the night supervisor in charge. After receiving four calls, he went outside and looked through his binoculars in the direction people had reported. He saw it, too—a bright object flashing colored lights, and changing positions at about 6,000 feet up. It also appeared to be changing shapes.

Pirouzi knew there were no planes or helicopters in the vicinity that night. At around 12:30 a.m., he alerted the Air Force command post. Deputy General Yousefi, who was in charge at the time, walked outside, and he also saw the object. He decided to scramble an Air Force Phantom F-4 II jet from Shahrokhi air base, located outside of Tehran, to investigate. The F-4 carried two people, Captain Aziz Khani and First Lieutenant Hossein Shokri, the navigator.

I was a major and the squadron commander at the time, and one of my pilots, who was among the first men alerted in the area, took off immediately. I left my house and headed for the base in order to be responsive to the operation there.

The F-4 was up when I arrived at the base, and both Khani and Shokri had seen the object and were attempting to chase it. But it was moving close to the speed of sound, so they couldn't catch it. When they came within a closer distance to it, all of their instrumentation went out, the radio was garbled, and they lost communication. After the F-4 moved away again, it regained all the instruments and could resume communications.

About ten minutes later, I was ordered to take off in a second jet to approach the object, which I was piloting. It was now about 1:30 a.m. on September 19. First Lieutenant Jalal Damirian, my second pilot in the backseat, operated the radar and other equipment; we called him "the backseater." When we took off, the object looked just like what had been reported. It was so brilliant, flying at a low altitude over the city, and then it started climbing.

Captain Khani had approached the Russian border, and at that point he was told to turn back. When he turned around, he said that he could see the object in front of him at twelve o'clock. I said, "Where exactly do you see it?" He said, "Over the dam, close to Tehran." I told him, "You go home, I'll take care of it." As he headed back, I looked over, and then I saw it.

It was flashing with intense red, green, orange, and blue lights so bright that I was not able to see its body. The lights formed a diamond shape—just brilliant lights, no solid structure could be seen through or around them. The sequence of flashes was extremely fast, like a strobe light. Maybe the lights were only one part of a bigger object, which we couldn't see. There was no way to know.

I approached, and I got close to it, maybe seventy miles or so in a climb situation. All of a sudden, it jumped about 10 degrees to the right. In an instant! Ten degrees . . . and then again it jumped 10 degrees, and then again. . . . I had to turn 98 degrees to the right from my heading of 70 degrees, so we changed position 168 degrees toward the south of the capital city.

I asked the tower whether they had it on radar. The operator replied, "The radar is out of order. It's not operational right now." All of a sudden my backseater, Lieutenant Damirian, said, "Sir, I have it on radar." I looked on the radar screen and saw the marker. I said, "Okay, brake lock and repaint it." This was to make sure it wasn't a ground effect or a mountain that we were picking up on the radar. We now had a good return on the screen, and it was at 27 miles, 30 degrees left; our closing speed was 150 knots and in a climb.

We kept it locked on with radar. The size on the radar scope was comparable to that of a 707 tanker.

At this moment, I thought this was my chance to fire at it. But when it—whatever it was—was close to me, my weapons jammed and my radio communications were garbled. We got closer, to 25 miles at our twelve o'clock position. All of a sudden it jumped back to 27 miles in an instant. I wondered what it was. I was still seeing that giant, brilliant diamond shape with pulsating, colored lights.

Then I was startled by a round object which came out of the primary object and started coming straight toward me at a high rate of speed, almost as if it were a missile. Imagine a brightly lit moon coming out over the horizon—that's what it looked like. I was really scared, because I thought that maybe they had launched some kind of projectile toward me. I had eight missiles on board, four operated by radar and four heat-seeking ones. The radar was locked on to the larger, diamond object, and I had to make a very fast decision as to what to do. I realized that if this moonlike, second thing *was* a missile, it would have some heat associated with it. So I selected an AIM-9 heat-seeking missile to fire at it.

I attempted to fire, and looked at the panel to confirm my selection of the missile. Suddenly, nothing was working. The weapons control panel was out, and I lost all the instruments, and the radio. The indicator dials were spinning around randomly, and the instruments were fluctuating. At this point, I was even more frightened. I couldn't communicate with the tower, and had to scream to talk to my backseater. I thought, if it gets closer to me than four miles, I will have to eject before impact to avoid being in the area of the explosion. To prevent this, I had to turn.

So I made a shallow turn to the left to avoid being impacted by the object heading toward us, which was in sight at my four-o'clock position. It came about four or five miles from our aircraft, and then it stopped there at the four-o'clock position. I looked out on my left side briefly to find out where I was over the ground. A second later, when I looked back, the object wasn't there! I said, "Oh my God," and Lieutenant Damirian replied, "Sir, it's at seven o'clock." I looked back at seven o'clock and there it was. I once again saw the main thing up there, too, and then the smaller object flew gently underneath it and rejoined the primary one.

This all happened quickly, and I didn't know what to think. But in a few seconds, another one came out! It started circling around us. Once again, all the instruments went out and the radio was garbled. Then, when it moved away, everything became operational again, and all the equipment worked fine. This one, too, looked sort of like the moon—a round, bright light.

I reported to the tower. General Yousefi was listening on the line, and the operator said, "The order is to come back." We started to head toward the military air base, and then I noticed that one of these objects was following us on our left side during the descent. I reported this to the base. As I made a turn for the final approach, I saw another object right ahead of me. I called the tower and asked, "I have traffic ahead of me, what is it?" He said, "We have no traffic." I said, "I am looking at it right now; it's at my twelve-o'clock position at a low altitude." He still insisted that I didn't have any traffic, but there it was, looking like a thin rectangle with a light at each end and one in the middle. It was coming toward me, but when I started turning left for the landing, I lost sight of it. My backseater kept watching and said, "As you were turning, I could see a round dome over it with a dim light inside of it."

I put the ears down and was focused on making my approach to the base, distracted and worried by all these things happening around me. But it still wasn't over. I looked to my left side and I saw the primary, diamond-shaped thing up there, and another bright object came out of it and headed directly toward the ground. I thought I would see a huge explosion any moment when it hit, but that did not happen. It seemed to slow down and land gently on the ground, radiating a high bright light, so bright that I could see the sands on the ground from that far, about fifteen miles.

I reported it to the tower and they said that they saw it, too. Now the general, still listening in, ordered me to approach and take a look. So I retracted the gear and the flaps and turned the aircraft. They told me to go above it to see if I could see what it was. As soon as I got about four or five miles from it, once again the radio was garbled and the panel went out; it was the same exact thing all over again. I tried to get out of that area because they couldn't hear me on the radio, and I told them, "This happens every time I get close to these things." I thought I really shouldn't have gone there, but since it was an order, I did it. Finally the general said, "Okay, come back and land."

We could hear emergency squawk coming from the location where the object had landed on the ground. A squawk sounds like the beeping from an ambulance or a police car, and its purpose is to help find people when they have ejected from an airplane, or if there is a crash landing. It's a locator tone that says "I'm here." In this case, the squawking from the UFO was reported by some civil airliners nearby.

After landing, I went to the command post, and then we went to check in with the tower. They said the main thing in the sky had just disappeared, suddenly, in an instant.

First thing that morning, I gave a report at headquarters, and everybody was in the room, all the generals. During this, an American colonel, Olin Mooy, a U.S. Air Force officer with the U.S. Military Advisory and Assistance Group posted in Tehran, sat to my left, and he was turning pages over on his clipboard and taking notes. When I explained how I couldn't fire the missile because my panel went out, even though I tried, he said, "You're lucky you couldn't fire." Afterward, I wanted to talk to him, and ask if this kind of thing had been seen before, and I had other questions. I looked for him, but he was nowhere to be found.

Next they then took me and Lieutenant Damirian to the hospital. We had a round of tests, especially blood tests. When I was about to leave, a doctor came running after me and said, "Don't worry about this, but your blood is not coagulating." So they took another blood sample, and then said, "Okay, you can go." They ordered us to return to the hospital every month for four months for an examination and more blood tests.

I then flew in a helicopter with a pilot and toured the exact area where the bright object had landed. The emergency squawk came from this area, and we flew right over the spot, but there was nothing. Nothing. We landed there, and I walked around to see if there was any sign of heating or burning, or splashing. Still nothing. Everything was smooth and untouched. Yet despite all that, the beeping was sounding. This was very confusing to us.

There were some small houses and gardens nearby and we asked the residents if they had seen anything. People said they had heard a sound the previous night after midnight, but that was it. The emergency squawk continued for days, and it was heard by the commercial airlines in the area, too. That really bothered me.

A group of scientists questioned us over a period of time, but it was all on paper, in letters sent to headquarters, and not in person. They called me in repeatedly from the base and I would go to headquarters and read the papers and answer more questions, again and again. Iranian officials examined and tested the two F-4s for radioactivity, and found none.

Later, a once-classified memo from the Defense Intelligence Agency (DIA), written by Lieutenant Colonel Mooy, whom I had tried to find after the briefing, was released in America through the Freedom of Information Act. It documented the event in great detail, for over three pages, and it was sent to the NSA, the White House, and the CIA. Another document, dated October 12, 1976, by Major Colonel Roland Evans, provided an assessment of the case for the DIA. It said that "This case is a classic which meets all the criteria necessary for a valid study of the UFO phenomenon."

To make that point, Evans listed some important facts in his DIA document: There were multiple highly credible witnesses to the objects from different locations; the objects were confirmed on radar; the loss of all instruments happened on three separate aircraft—a commercial jet as well as our two F-4s; and "an inordinate amount of maneuverability was displayed by the UFOs." The evaluation form said that the reliability of the information was "confirmed by other sources" and the value of the information was "high." It said the information would be potentially useful. This shows the U.S. government took this information very seriously, and it was clear to me at the time that this information was being kept secret there. But within a relatively short time these documents were released. There is likely additional material sitting in U.S. government files, but no one has told me anything more.

In my country, even the Shah of Iran took an interest. I met with the shah when he visited my squadron at Shahrokhi air base in Hamadan and asked about the UFO. He called a meeting attended by a number of generals along with the pilots involved in the encounter. When the base commander told the shah that I was the pilot who had chased the UFO, the shah asked me, "What do you think about it?" I answered, "In my opinion they can not be from our planet, because if anyone on this planet had such power, he would bring the whole planet under his own command." He simply said, "Yes," and told us this was not the first report he had received.

To this day I don't know what I saw. But for sure it was not an aircraft; it was not a flying object that human beings on Earth can make. It moved way too fast. Imagine: I was looking at it about seventy miles out and it jumped all of a sudden 10 degrees to my right. This 10 degrees represented about 6.7 miles per moment, and I don't say per second because it was much less than a second. Now you can try to calculate the speed it would take for it to move from a stationary position to this second point. This needed very, very high-level technology. Also, it was able to shut down my missile and instruments somehow. Where it came from, I don't know.

And I can't doubt what happened. It wasn't only me. The pilot in my backseat, the two pilots in the first aircraft, the men in the tower, people from headquarters, General Yousefi who was on duty in the Air Force command post—they all saw it. Many people were concerned about us on the ground. And we also captured it on radar from our cockpit. Nobody can say I imagined it. The radar was locked on the object and could determine its size, because we practice refueling 707 tankers, and the return of the UFO on radar indicated they were about the same size.

I have two regrets: One is that we did not have a camera in the plane to get a picture of the UFO; second, that because I was excited and sometimes frightened, I didn't think to try and call them on the radio, and ask, "Who are you? Please communicate with us!" Later on I wished I had done this. In any case, I hope someday we develop that technology here so we can travel easily to other planets and poke around, too.

CHAPTER 10

Close Combat with a UFO

by Comandante Oscar Santa María Huertas (Ret.), Peruvian Air Force

On April 11, 1980, at 7:15 a.m., a Friday morning, I was stationed at the La Joya Air Force Base in the Arequipa region of Peru. It was like any other day. There were approximately 1,800 military personnel and civilians at the base, and we were beginning to get ready for our daily exercises.

Even though I was only a twenty-three-year-old lieutenant, I already had eight years of military flying experience. I was quite precocious as a military pilot. By nineteen I was flying combat missions, and at twenty I was selected to test-fly Peru's newest supersonic Sukhoi jet. Having won quite a few trophies as a pilot, I was also known as a top aerial marksman with great skill at shooting from the air.

Little did I know that this expertise would lead to my being selected for a highly unusual and unexpected mission on that routine morning. Along with my air squadron, I was ready at that moment for instant take-off, as we always are. A chief of service arrived in a van and got out to tell us there was an object that looked like some kind of balloon suspended in the air toward the end of the runway. We stepped outside to see it, and then we knew what we had to do. Four of us pilots stood outside observing the object. The second commander of the unit, Commander FAP Carlos Vasquez Zegarra, ordered that one of the members of the air squad take off and bring the object down. Our chief turned to me and said, "Oscar, you be the one to go."

The round object was about three miles (five kilometers) away from us, hanging at an altitude of about 2,000 feet (600 meters) above the ground. Since the sky was absolutely clear, the object shone due to the reflection of the sun.

This "balloon" was in restricted air space without authorization, representing a grave challenge to national sovereignty. All civilian and military pilots use aerial charts on which highly protected airspace, such as that over our base, is clearly marked. They all know where these restricted areas are located, and no one ever flies in them, under any circumstances. This thing had not only appeared in such an area, but it was not replying to communications sent on universally recognized frequencies, and it was moving toward the base. It had to come down. La Joya was one of the few bases in South America that possessed Soviet-made warfare equipment, and we were concerned about espionage.

Back in 1980, Peru did not have any aerostatic balloons of any type, such as weather balloons, or passenger balloons. We knew that this was therefore something strange, and it wasn't from our country. We were familiar with meteorological balloons, but they had antennae and cables and flew only above 45,000 feet. This was lower. We had no idea where it was from, and it was coming closer. We had no option but to destroy it.

The squad commander, Captain Oscar Alegre Valdez, ordered me to take off in my Sukhoi-22 fighter jet to intercept the balloon before it got any closer to our base. I immediately headed over to my jet, without taking my eyes off the thing in the sky, and went over in my mind each step I had to take for this mission. Since the object was within the perimeter of the base and my plane was armed with 30 mm shells, I decided to attack from the northeast to the southeast. This way, the sun would be to my left and I could avoid impacting the base with my weapons.

After takeoff, I made a turn to the right and reached an altitude of 8,000 feet (2,500 meters). I then positioned myself for the attack. Zeroing in on the balloon, I reached the necessary distance and shot a burst of sixty-four 30 mm shells, which created a cone-shaped "wall of fire" that would normally obliterate anything in its path. Some of the projectiles deviated from the target, falling to the ground, and others hit it with precision. I thought that the balloon would then be torn open and gases would start pouring out of it. But nothing happened. It seemed as if the huge bullets were absorbed by the balloon, and it wasn't damaged at all. Then suddenly the object began to ascend very rapidly and head away from the base.

"What is going on here?" I thought to myself. "I have to get closer to it."

So I headed up. I initiated a chase by activating the afterburner of my plane, and reported to the control tower that I intended to follow procedures and continue the task of bringing down the object. Since I knew that this was an extremely unusual mission, I asked that they make sure the tape recorders were working so that anything taking place from that moment on would be on record. Then, an amazing series of events unfolded.

My jet flew at a speed of 600 mph (950 km/hr) and the "balloon" remained about 1,600 feet (500 meters) in front of me. As we got farther from the base, I reported to the control tower information such as "I am at three thousand meters of altitude and twenty kilometers from the base . . . I am at six thousand meters of altitude and forty kilometers from the base . . ." and so on. By this time I was over the city of Camana, which was about fifty-two miles (eighty-four kilometers) from the base, flying at 36,000 feet (11,000 meters).

I was in full pursuit of the object, when it came to a sudden stop and forced me to veer to the side. I made a turn upward to the right and tried to position myself for another shot. Once I obtained the desired position to fire, which was approximately 3,000 feet (1,000 meters) from the object, I began closing in on it until I had it in perfect sight. I locked on the target and was ready to shoot. But just at that moment, the object made another fast climb, evading the attack. I was left underneath it; it "broke the attack."

I attempted this same attack maneuver two more times. Each time, I had the object on target when it was stationary. And each time, the object escaped by ascending vertically seconds before I started to fire. It eluded my attack three times, each time at the very last moment.

Throughout this time I was very focused on trying to achieve my window of about 1,300 to 2,300 feet (400 to 700 meters) distance, which was where I had to position my plane in order to shoot. As this became less possible, I was very surprised and kept asking myself what was going on. Then it became a personal thing for me. I *had* to get it. But I couldn't because it would always ascend. I was committed to this mission, and felt I must succeed. This was all that mattered, and I felt confident knowing I had an outstanding airplane.

Eventually, as a result of this series of rapid movements upward, the object ended up at an altitude of 46,000 feet (14,000 meters). I had to think of something else to do! I decided to make a bold ascent with my plane so that this time I would be *above* the object, and then I would come down on it vertically and initiate an attack from above. This way, if the object began to ascend as in the previous three attempts, it would not leave my target range and it would be easier for me to fire. I was not concerned about any collision because of the agility and maneuverability of my plane.

So I accelerated my plane at supersonic speed and went back to where the "balloon" was, by this time traveling at a speed of Mach 1.6, which is approximately 1,150 mph (1,850 km/hr). I calculated the distance between the object and myself as I began to make the ascent. As I went higher, I saw that the object was in fact under me and I thought I would be able to gain the necessary altitude to pull off the maneuver as planned, and succeed in the attack. But to my surprise, the object ascended once again at a high speed and placed itself next to me in parallel formation! This left me without any possibility of attack.

Flying at Mach 1.2, I continued with my ascent, still hoping to pass above it in order to initiate the attack I had planned. But I couldn't. We reached an altitude of 63,000 feet (19,200 meters or 19 kilometers), and suddenly the thing completely stopped and remained stationary. I adjusted the wings of my plane to 30 degrees and extended its slats so that the plane would be able to maneuver at that height, and I thought I could still attempt to target the object in order to fire. But it was impossible. I could not remain as still as this "balloon."

At that moment the warning light for low fuel went off, indicating that I had just enough to get back to the base. Under those conditions, I could not continue the attack, so I flew closer to the hovering object to observe it and try to determine what it was. The Su-22s had no on-board radar, but the sighting equipment had well-marked gradations that communicated the distance from a target and its diameter. This technology was based on the use of laser beams.

I got as close as about 300 feet (100 meters) from it. I was startled to see that the "balloon" was not a balloon at all. It was an object that measured about 35 feet (10 meters) in diameter with a shiny dome on top that was cream-colored, similar to a light bulb cut in half. The bottom was a wider circular base, a silver color, and looked like some kind of metal. It lacked all the typical components of aircraft. It had no wings, propulsion jets, exhausts, windows, antennae, and so forth. It had no visible propulsion system.

At that moment, I realized this was not a spying device but a UFO, something totally unknown. I was almost out of fuel, so I couldn't attack or maneuver my plane, or make a high-speed escape. Suddenly, I was afraid. I thought I might be finished.

After recovering from the impact of seeing this, I began my return to the air base and explained to the control tower exactly what I had seen. When I had calmed down, I radioed for another plane to come and continue the attack, trying to hide my fear. They said no, it's too high, just come back. I had to glide partway down due to lack of fuel, zigzagging to make my plane harder to hit, always with my eyes on the rearview mirrors, hoping it wouldn't chase me. It didn't. I had been flying for twenty-two minutes.

As I was touching down, I was very excited and couldn't wait to tell my people about the extraordinary thing I had flown against. It was so fascinating that I had really wanted someone else to come up and take a look. I had described this object as flying, even though it had no visible equipment for that—nothing to make it fly!

When I stepped out of my plane, my squadron was waiting and asked me lots of questions. The maintenance person was there and checked the shell cartridges and said, "Captain, it's clear you've done some shooting." Others came by, and there were many inquiries and conversations.

Right after my landing, all the personnel involved in the incident gathered for a briefing—this meant operations personnel, air defense, base defense, and the general who was the aerial wing commander. Due to the threat established by this "balloon," our base had activated its defense system and all systems were on alert. Everyone gave reports. We learned that the object was never registered on radar, even though the radar operators could see it in the sky, as could the people who had observed the object early on when it was stationary. They also described it as round and metallic. We were told that what happened at the meeting was to remain there only, and we were not to divulge it at any time.

After this briefing, I met with intelligence personnel and we went over all the catalogues available with pictures of different types of aircraft or air devices employed for espionage, but we found nothing that could maneuver in the way I described without any type of propulsion system. The object was consequently catalogued as an unidentified flying object. It remained in the same place where I left it for two more hours, visible to everyone on the base while it reflected the sun.

I never saw any U.S. government officials at the base discussing this case, and they never interviewed me. Nonetheless, a document of the Defense Department of the United States dated June 3, 1980, titled "UFO Sighted in Peru," describes the incident and states that the object remains of unknown origin.

In conclusion, I can say that in 1980 I had a combat experience with an unidentified flying object that flew and maneuvered in the air without any recognizable features of aircraft, features that even today are necessary parts of any flying machine. This object performed maneuvers that defied the laws of aerodynamics. After thorough investigation of all available data regarding aircrafts, our military experts were not able to find any artifact or machine that could have done what this object did.

Many years later, I have learned of similar cases in which military planes have chased unidentified flying objects without being able to successfully launch their weapons due to the fact that their systems were blocked prior to firing. I have discussed this with experts from around the world, including those at the National Press Club event in Washington, D.C., in November 2007. Both the Iranian case of 1976 and another similar case in Brazil involved the shutting down of *electronic* equipment—the control screens went out. My equipment was mechanical, and perhaps that's the reason it could not be shut down, so instead, the object had to jump away at the last minute.

I find myself in the unique position, at least for the moment, and as far as I know, of being the only military pilot in the world who has actually fired a weapon and struck a UFO.

It still gives me chills to think about it.

PART 2

IN THE LINE OF DUTY

"It is one of the ironies of modern rule that it is far more acceptable today to affirm publicly one's belief in God, for whose existence there is no scientific evidence, than UFOs, the existence of which—whatever they might be—is physically documented."

ALEXANDER WENDT AND RAYMOND DUVALL

CHAPTER 11

The Roots of UFO Debunking in America

B ecause all of us have long been exposed to an atmosphere of
ridicule and the automatic dismissal of the UFO phenomenon,
I suspect that the information presented so far may have been
very surprising, even shocking, for some readers. It's not easy for anyone
to come to terms with evidence for the reality of UFOs, and yet we've
seen that such evidence can't be dismissed out of hand. In reading about
General De Brouwer's painstaking investigation, or the disc hovering
above O'Hare Airport, or the huge, flashing object jumping through the
night sky over Tehran, we find ourselves forced to reconcile two radically
conflicting paradigms. There is the one position we've always known, in
which these things are out of the question; they can't happen, according to agreed-upon laws of physics and cosmology, and therefore they
simply *don't* happen. But then there's the fact that unknown objects have
been seen by thousands of people all over the world, demonstrating
these "impossible" capabilities right before our eyes. Most disturbing, of
course, is the implied possibility that these UFOs, apparently under some
kind of intelligent control, *might* have an origin outside of planet Earth,
no matter how unthinkable that idea may be.

The reader may feel bewildered by this possibility, incredulous and
hesitant to go on. There may still be that inclination to dismiss it all as
foolishness or some kind of psychological aberration that no amount of
evidence can change. Some readers might feel defiant at this point, or
deeply alarmed. Simple curiosity and an open mind will temper these
very natural reactions. Anyone who adventures into this strange realm
goes through some level of internal struggle, as I did after discovering
and researching the COMETA Report. Like everyone else, I was unnerved
by all this, but also, as an investigative journalist, I soon became intrigued

by its power and portentousness. As I've already described, I wanted to find out as much as I could about the UFO phenomenon—*really* find out if there was anything to it. And after a while I developed a kind of defiance—but this was not because of resistance to accepting UFOs as real. Instead, I was disturbed that something real *was* going on here and nobody seemed to be paying attention. Being naturally rebellious, I felt drawn to the challenge both to my own intellectual boundaries and to the limitations of conventional thinking. Awe and humility softened the more unnerving aspects of the discovery process, because the more I learned, the more convincing the whole thing became. Why should we assume we already understand everything there is to know, in our infancy here on this planet?

My evolution took years, involving much reading, discussion with veteran researchers, review of government documents, and interviews with retired military officials and UFO witnesses. I think that most of us willing to consider this subject, even without this level of intensity, come to a point of transition, a decisive moment when we cross our own deeply ingrained internal barrier. It isn't easy. After all, we're dealing with something so far ungraspable: the essential nature of the UFO. We have to come to terms with the recurring appearance of something absolutely unknown and unexplainable by science, something that operates as if it were outside the boundaries of our physical world but *in it* at the same time. To make it even more difficult, we're burdened by the negativity and denial of the status quo that all of us have absorbed to one degree or another.

To understand that aspect of the problem, we must come down to earth and look at the political and historical roots of the U.S. government's reaction to the UFO phenomenon, beginning at a time when officials first recognized that they were dealing with something not easily explained. Even if the phenomenon is psychologically hard to confront, that excuse is not enough to explain the inaction, the dismissal, and the ridicule that have been the norm for so many years. Why is there such a strong taboo against taking the subject seriously, when there's so much evidence for it?

In fact, our government has a policy—a stated position of inaction crafted over fifty years ago—underlying its current approach to UFOs. Certain pivotal events set us on the unfortunate course we find ourselves on today. It all began in the late 1940s, when officials were faced with a sudden influx of UFO sightings in the skies over America, many of which were reported by highly credible observers such as military and air-line pilots. Popular interest in UFOs (called "flying saucers" at the time, because of their frequently described flattened-disc shape) was growing as a result of national media coverage and the fact that nobody knew what they were or how to handle them. Initially, the authorities attempted to determine if the objects were either secret foreign aircraft, such as supe-rior technology from the Soviet Union, or possibly some kind of newly discovered atmospheric or meteorological phenomena.

By 1947, things were becoming uncomfortably clear behind the scenes. Lieutenant General Nathan Twining, commander of Air Force Materiel Command, a major command of the U.S. Air Force, sent a secret memo concerning "Flying Discs" to the commanding general of the Army Air Forces at the Pentagon. The considered opinion, based on data furnished by numerous Air Force branches, he stated, was that "the phenomenon reported is something real and not visionary or fic-titious . . . The reported operating characteristics such as extreme rates of climb, maneuverability (particularly in roll), and action which must be considered evasive when sighted or contacted by friendly aircraft and radar, lend belief to the possibility that some of the objects are con-trolled either manually, automatically or remotely." Twining described the objects as metallic or light-reflecting, circular or elliptical with a flat bottom and domed top, sometimes with "well kept formation lights vary-ing from three to nine objects," and normally silent. He proposed that the Army Air Forces set up a detailed study of UFOs, assigning a security classification and code name to it.

As a result, such a project was set up within the Air Materiel Com-mand, and given the code name "Sign." The new agency began its oper-ations in early 1948 at Wright Field (now called Wright-Patterson Air Force Base) with the mandate to collect information, evaluate it, and assess whether the phenomenon was a threat to national security. As Project Sign became more convinced that the objects were not Russian, divisions grew between those who thought they were "interplanetary"— the term used at the time, when much less was known about our solar system—and those who were determined to find a more conventional explanation. Later that year, some Project Sign staff wrote a top-secret report, an "Estimate of the Situation," providing data on convincing cases and concluding that, based on the evidence, UFOs were most likely extraterrestrial. The document eventually landed on the desk of General Hoyt Vanderbeng, Air Force Chief of Staff, who rejected it as unacceptable because he wanted proof, and responded by returning it to its authors at Project Sign. From then on, the proponents of the extraterrestrial hypothesis lost ground, and because of the clear message from Vandenberg and others, the safer position that UFOs *must* have conventional explanations was adopted by the majority of the project's investigators. It appears they were under pressure to shift their focus. The "Estimate of the Situation" was reportedly destroyed, and no copies have ever been found despite repeated attempts using the Freedom of Information Act.

Project Sign was later renamed Project Grudge, which then became the well-known Project Blue Book in 1951, lasting for nineteen years. As time passed, it continued to become naggingly clear that these objects did *not* belong to any foreign government, and we had to face the clear possibility that they did not originate here on Earth. U.S. government documents released through the FOIA show that, as a result, some officials from multiple branches of government continued to assert that they might be interplanetary. As before, other factions stuck to their hope of finding a conventional explanation, no matter what.

In July 1952, the FBI was briefed through the office of Major General John Samford, the director of intelligence for the Air Force, and told that it was "not entirely impossible that the objects sighted may possibly be ships from another planet such as Mars." Air intelligence was "fairly certain" that they were not "ships or missiles from another nation in this world," the FBI memo reports. Another FBI memo stated some months later that "some military officials are seriously considering the possibility of planetary ships."

At the same time, national defense concerns were mounting about the preponderance of technologically advanced unidentified objects flying over the United States during the Cold War. One famous series of sightings over the nation's capitol, in which Air Force planes were sent to intercept brilliant objects picked up by ground radar, made national headlines in July 1952, and necessitated a press conference, the biggest one since World War II, in which intelligence chief General Samford tried to calm the country. He said:

Air Force interest in the problem has been due to our feeling of an obligation to identify and analyze, to the best of our ability, anything in the air that has the possibility of [being] a threat or menace to the United States. In pursuit of this obligation, since 1947, we have received and analyzed between one and two thousand reports that have come to us from all kinds of sources. Of this great mass of reports, we have been able adequately to explain the great bulk of them—explain them to our own satisfaction. However, there are then a certain percentage of this volume of reports that have been made by credible observers of relatively incredible things. It is this group of observations that we now are attempting to resolve. We have, as of date, come to only one firm conclusion with respect to this remaining percentage. And that is that it does not contain any pattern of purpose or of consistency that we can relate to any conceivable threat to the United States.

He told reporters that the Washington, D.C., events were likely mere aberrations caused by temperature inversions—layers in the atmosphere in which rising temperatures affect radar performance—an interpretation disputed by the pilots and radar operators involved.

The increasing numbers of reports were becoming hard to manage along with growing public interest in the phenomenon. In late 1952, H. Marshall Chadwell, assistant director of scientific intelligence for the CIA, sent a memo about this problem to the Director of Central Intelligence (DCI). "Sightings of unexplained objects at great altitudes and travelling at high speeds in the vicinity of major U.S. defense installations are of such nature that they are not attributable to natural phenomena or known types of aerial vehicles," he stated.

In another 1952 memo, titled "Flying Saucers," the CIA's Chadwell said the DCI must be "empowered" to initiate the research necessary "to solve the problem of instant positive identification of unidentified flying objects." The CIA recognized the need for a "national policy" as to "what should be told the public regarding the phenomenon, in order to minimize risk of panic," according to government documents. It was therefore decided that the DCI would "enlist the services of selected scientists to review and appraise the available evidence." As a result of this decision, the CIA arranged a critically important meeting that would forever change both the course of media coverage and the official attitude toward the UFO subject. The results of this meeting help explain the omnipresent disengagement of American officials during the decades to come.

The CIA began its work in January 1953, when it convened a hand-picked scientific advisory panel, chaired by H. P. Robertson, a specialist in physics and weapons systems from the California Institute of Technology, for a four-day closed-door session. Authorities were concerned that communication channels were being so saturated by hundreds of UFO reports that they were becoming dangerously clogged. Even though the UFOs had demonstrated no threat to national security, false alarms could be dangerous and defense agencies might have a problem discerning true hostile intent. Officials were concerned that the Soviets might take advantage of this situation by simulating or staging a UFO wave, and then attack.

Thus the Robertson Panel's goal was to find ways *to reduce public interest* in order to prevent the filing of reports. Members of the distinguished panel were given a cursory review of selected UFO cases and exceptional film footage that had so far been kept secret. This was meant to represent an overview of the best UFO data on file, but the four days allotted was not nearly enough time for a proper assessment. Nonetheless, in its secret report written at the completion of its review, the Robertson Panel recommended that "the national security agencies take immediate steps to strip the Unidentified Flying Objects of the special status they have been given and the aura of mystery they have unfortunately acquired."

How would they achieve this? The panel proposed the creation of a broad educational program integrating the efforts of all concerned agencies, with two major aims: training and debunking. Training meant more public education on how to identify known objects in the sky, so that they would not be misidentified as UFOs. Debunking was for use primarily by the media. "The 'debunking' aim would result in reduction in public interest in 'flying saucers' which today evokes a strong psychological reaction," wrote the panel, "and would be accomplished by mass media such as television, motion pictures, and popular articles."

In addition to the media, the panel recommended using psychologists, advertising experts, amateur astronomers, and even Disney cartoons to reduce enthusiasm and gullibility. "Business clubs, high schools, colleges, and television stations would all be pleased to cooperate in the showing of documentary type motion pictures if prepared in an interesting manner. The use of true cases showing first the 'mystery' and then the 'explanation' would be forceful." Lastly, civilian groups studying UFOs should be "watched" due to their "great influence on mass thinking if widespread sightings should occur."

In short, a group of scientists selected by the CIA advised our government to encourage all agencies within the intelligence community to influence mass media and infiltrate civilian research groups for the purpose of debunking UFOs. Media could then become a tool for covertly controlling public perception, a mouthpiece for government policy and propaganda, to "debunk," or ridicule, UFOs. Public interest in UFO incidents was to be strongly discouraged and diminished through these tactics, and intelligence operatives could make sure that the facts were kept from leading researchers through disinformation. In the name of national security, the subject was fair game for the entire U.S. intelligence apparatus. All of these recommendations were written in black and white by the CIA panel and then classified, and the public did not have access to the full report until 1975, when the explosive Robertson Panel Report was finally released in its entirety.

When the CIA convened its selected group of scientists in 1953, astronomer J. Allen Hynek had been working for a number of years as consultant to the U.S. Air Force's Project Blue Book. Formerly director of Ohio State University's McMillan Observatory and later chairman of the astronomy department and director of the Lindheimer Astronomical Research Center at Northwestern University, Dr. Hynek had been hired in 1948. He sat in on most of the Robertson Panel meetings and observed the predetermined agenda unfold, noting that the best UFO evidence was not given proper attention. "The implication in the Panel Report was that UFOs were a nonsense (non-science) matter, to be debunked at all costs," Hynek revealed later. "It made the subject of UFOs scientifically unrespectable."

Project Blue Book had been set up as a repository for UFO cases and a place for people to call and file reports of sightings, but in reality it was an understaffed, amateurish public relations operation focused on explaining away UFO sightings, no matter how far-fetched the explanation.

Throughout his career as popular public representative of Blue Book for the duration of its operation, Hynek was well aware of the integration of the "training and debunking" tactic within the Air Force program, but ironically, as one of the implementers of the Robertson Panel agenda, he was part of the problem himself.

Years later he admitted that "for nearly twenty years [of Project Blue Book, 1951–1970] not enough attention was paid to the subject to acquire the kind of data needed even to decide the nature of the UFO phenomenon." Hynek was the only consistent presence at Blue Book and the sole scientist. The office was staffed mainly by an ever-changing stream of low-ranking officers with no particular training to prepare them for this line of work, and often little interest in it. Hynek brought some respectability to the Air Force project, though it was never equipped to solve the problem and official prejudice kept it that way.

Despite his eventual transformation after two decades of work with the Air Force, Hynek had earlier stretched logic to its limit in order to explain away as many UFO reports as possible. In his landmark 1972 book *The UFO Experience: A Scientific Inquiry,* he acknowledged that debunking was what the Air Force expected of him. "The entire Blue Book operation was a foul-up based on the categorical premise that the incredible things reported could not possibly have any basis in fact," he wrote. The Air Force, at least publicly, had dutifully fulfilled the debunking role that the CIA panel had so highly recommended, and Blue Book records are rife with examples of solid cases being given ridiculous, often infuriating explanations, sometimes by Hynek himself. Even as he became more aware of the contradiction in later years, Hynek said he did not want to fight with the military and felt it was more important that he maintain access to the store of data at Blue Book, "as poor as they were."

In this vein, perhaps most famous is his "swamp gas" statement, made in 1966. For two days, over a hundred witnesses in Dexter and Hillsdale, Michigan, had seen glowing unidentified objects at relatively low altitudes, many of them near swampy areas. This quickly became a highly charged national news story, and great pressure was placed on the Air Force to solve the case as quickly as possible. Hynek was called to a packed press conference, one bordering on hysteria, as he described it, where he made the comment that the lights could have been the glow of something called marsh gas, a rare phenomenon that arises from the spontaneous ignition of decaying vegetation. The hostility he faced in the press and among the public for his "swamp gas" explanation was widespread, and the media ridicule he received is now legendary. This time, everyone seemed to recognize that the Air Force had gone too far and crossed an unacceptable line in its debunking.

American frustration with the Air Force's inability to adequately investigate and address recurring UFO sightings had been building, and many now began to feel that the Air Force was not only incompetent but actually intent on covering up the truth about UFOs. Two well-known figures of this era—Major Donald Kehoe of the National Investigations Committee on Aerial Phenomena, a leading civilian research group, and Dr. James E. McDonald, a senior atmospheric physicist from the University of Arizona—played critical roles in bringing credibility and knowledge to the UFO subject while challenging the approach of Project Blue Book. Following the publication of best-selling books and magazine cover stories about UFOs that year, public interest in the phenomenon was at its peak.

We will never know to what extent the recommendations of the Robertson Panel were directly implemented, but we do know that one of the Robertson panelists stepped up to the plate in 1966. Astrophysicist Thornton Page of Johns Hopkins University wrote to Frederick Durant, head of the National Air and Space Museum's aeronautics department—both men had been members of the Robertson Panel—claiming that he "helped organize the CBS TV show around the Robertson Panel conclusions," referring to the two-hour special "UFO: Friend, Foe or Fantasy?" hosted by the trusted Walter Cronkite. The Cronkite show debunked UFOs from all angles with intense bias and false claims, such as statements that no radar or photographic evidence existed to support the physical reality of UFOs. It seems clear that *someone* must have been operating behind the scenes to justify such an extreme position. Ironically, Thornton Page himself made an appearance on the CBS special, defending the objectivity of the Robertson Panel evaluation and telling viewers that "we tried to evaluate all the reports without saying they're ridiculous in advance." Cronkite reported that the CIA panel found "no evidence of UFOs" and ended the broadcast by encouraging viewers to remember that "while fantasy improves science fiction, science is more served by fact."

Due to the outrage of his constituents following a series of sightings in his state, including the ones labeled "swamp gas," Representative Gerald Ford, House Republican minority leader at the time, "in the firm belief that the American public deserves a better explanation than thus far given by the Air Force," called for congressional hearings on the subject of UFOs. Just before the Cronkite special, on April 5, 1966, the House Armed Services Committee heard from members of the Air Force, including consultant J. Allen Hynek, about the UFO problem, in which they considered recommendations for an independent scientific investigation outside of Project Blue Book. The Air Force took its first step away from the messy UFO business by agreeing to find a university willing to coordinate the study, one which would help the Air Force decide whether to continue its own program or disentangle itself from an unsatisfactory public relations campaign becoming increasingly difficult to maintain.

Late in 1966 it was decided: The University of Colorado agreed to host a government-funded study of UFOs to be headed by Edward U. Condon, a well-known physicist and former head of the National Bureau of Standards. Although initial expectations were high for the project, and for a short time even added legitimacy to scientific scrutiny of UFOs, it gradually fell apart due to internal disputes among the study's committee members. It soon became known that from the outset Condon had held strongly negative personal views about the subject and had never intended to proceed fairly or objectively. On top of that, conflict arose about whether the extraterrestrial hypothesis had any validity along with the many other theories under consideration. A crisis point was reached when two concerned project members unearthed a damaging August 9, 1966, memo by project coordinator Robert Low to two university deans. In it, Low had discussed the pros and cons of taking on the UFO research project, when it was still under discussion.

If the project were to be undertaken, he laid out the problem:

> One has to approach it objectively. That is, one has to admit the possibility that such things as UFOs exist. It is not respectable to give serious consideration to such a possibility . . . one would have to go so far as to consider the possibility that saucers, if some of the observations are verified, behave according to a set of physical laws unknown to us. The simple act of admitting

these possibilities just as possibilities puts us beyond the pale, and we would lose more in prestige in the scientific community than we could possibly gain by undertaking the investigation.

So, Low offered a way out:

> Our study would be conducted almost exclusively by nonbelievers who, although they couldn't possibly *prove* a negative result, could and probably would add an impressive body of evidence that there is no reality to the observations. The trick would be, I think, to describe the project so that, to the public, it would appear a totally objective study but, to the scientific community, would present the image of a group of nonbelievers trying their best to be objective, but having an almost zero expectation of finding a saucer.

The specific language he used in his memo—particularly the word "trick"—helped give his game away. The term "flying saucer" was often used in conjunction with "believers" and "enthusiasts," who assumed the objects were extraterrestrial and were (presumably) not using the scientific method to address the problem. Condon was infuriated that this was made public, and he fired the two staffers who had leaked the memo the day after he heard about it.

Although Low attempted to keep his own views secret, Condon had no problem making his negative attitudes toward his subject public. In a January 1967 lecture he remarked, "It is my inclination right now to recommend that the government get out of this business. My attitude right now is that there's nothing to it." He added, "But I'm not supposed to reach a conclusion for another year."

In response to public concern about all of this, and in reaction to continuing dramatic UFO sightings, a second congressional hearing was called by the House Science and Astronautics Committee in July 1968. A host of scientists from outside the Air Force presented compelling papers on their own studies of UFOs; many of them had grave reservations about the effectiveness of the Condon study and advocated the continued study of UFOs despite its outcome. The testimony of Dr. James E. McDonald, from the Institute of Atmospheric Physics and a professor of meteorology at the University of Arizona, was the most extensive, providing a series of compelling UFO case reports. A respected authority and leader in the field of atmospheric physics, McDonald had written many highly technical papers for professional journals. Due to his personal interest, he spent two years examining formerly classified official file material and radar tracking data on UFOs, interviewing several hundred witnesses, and conducting in-depth case investigations on his own, details of which were provided to the committee.

McDonald testified that no other problem within their jurisdiction compared to this one. "The scientific community, not only in this country but throughout the world, has been casually ignoring as nonsense a matter of extraordinary scientific importance." He indicated that he leaned toward the extraterrestrial hypothesis as an explanation, due to "a process of elimination of other alternative hypotheses, not by arguments based on what I could call 'irrefutable proof.' " Dr. Hynek recommended that a congressional UFO scientific board of inquiry set up a mechanism for the proper study of UFOs, "using all methods available to modern science," and that international cooperation be sought through the United Nations.

Extensive research has been done and books have been written on the tumultuous process which eventually produced the Condon committee report, "Scientific Study of Unidentified Flying Objects," released in 1968. The approximately 1,000-page tome begins with the conclusions and recommendations by Condon himself. He declared that further scientific study of UFOs was unwarranted and recommended that the Air Force shut down Project Blue Book. Nothing should be done with UFO reports submitted to the federal government from then on, he believed. He wrote that no UFO has posed a national security or defense problem, and that there was no official secrecy concerning UFO reports. Condon's two-page summary of the report, released to the press and public, actually contradicted the findings contained within the body of the volume, which most people did not bother to read.

In fact, Condon himself did not participate in the analysis of the carefully researched case studies that made up the bulk of the study, and it appears he also didn't bother to read the finished product. The lengthy study *did* provide some excellent scientific analysis by other members of the committee, buried among many tedious case analyses of marginal importance which dragged on, page after page. Other key cases were left out altogether. Some reports actually verified the reality of still unsolved and highly perplexing UFO phenomena. For example, investigator William K. Hartman, astronomer from the University of Arizona, researched two extraordinary photographs from McMinnville, Oregon, and stated that "this is one of the few UFO reports in which all factors investigated, geometric, psychological, and physical, appear to be consistent with the assertion that an extraordinary flying object, silvery, metallic, disc-shaped, tens of meters in diameter, and evidently artificial, flew within the sight of two witnesses."

Regardless, Condon's summary stated, "Nothing has come from the study of UFOs in the past twenty years that has added to scientific knowledge." And the National Academy of Sciences endorsed Condon's recommendations. "A study of UFOs in general is not a promising way to expand scientific understanding of the phenomena," it concluded seven weeks later. Condon added insult to injury by telling the *New York Times* that his investigation "was a bunch of damn nonsense," and he was sorry he "got involved in such foolishness."

The American Institute of Aeronautics and Astronautics (AIAA) was among those registering objections after its panel spent over a year studying the actual 1,000-page text of the Condon report. The AIAA stated that Condon's summary did not reflect the report's conclusions but instead "discloses many of his [Condon's] personal conclusions." The AIAA scientists found no basis in the report for Condon's determination that further studies had no scientific value, but declared instead that "a phenomenon with such a high ratio of unexplained cases (about 30% in the Report itself) should arouse sufficient curiosity to continue its study."

Behind Condon's and Low's disdain and closed minds, along with those of others in that camp, lay, once again, the problem of confronting the extraterrestrial hypothesis. As Hynek pointed out at the time, Condon and his supporters mistakenly equated the notion of UFOs with something extraterrestrial, believing that if UFOs were acknowledged as a genuine phenomenon, an implicit acceptance of the extraterrestrial hypothesis would ensue. This was clearly unacceptable to them. As Low pointed out in his memo, the simple act of admitting such a possibility was "beyond the pale," and any professional doing so risked losing prestige within a scientific community not open to such a radical concept.

Even after twenty-two years of Air Force accumulation of data, along with independent studies made by various scientists such as McDonald, an overwhelming number of scientists and government officials still felt profound unease with entertaining even the remote possibility of such a hypothesis. That aversion was strong enough that its purveyors didn't mind that it completely undermined the accuracy and effectiveness of an expensive, years-long scientific study on which so much depended, and which everyone knew would have a huge, historical impact.

Instead, the final nail was in the coffin. In December 1969, the Air Force announced the termination of Project Blue Book—our government's only official investigation of UFOs—effective the following month. From then on, scientists could justify their dismissal of UFOs by citing the conclusions of the Condon report. The government could refer to the Air Force decision to end its investigation to justify its disinterest in UFO cases. The media could enjoy the ride while making fun of UFOs or relegating them to science fiction. Now, no more direct action was required by those carrying out the mission of the Robertson Panel because the seeds had all been planted and the momentum would be self-generating for decades to come. The "golden age" of official investigations, congressional hearings, press conferences, independent scientific study, powerful citizen groups, best-selling books, and magazine cover stories had come to an end.

In the decades following, many dedicated researchers carried the torch and devoted their lives to documenting cases and adding to our knowledge of the phenomenon. Their capable and extensive work has been crucial in carrying us forward. But once an issue galvanizing concern on the national stage, the UFO question now shifted to the margins. The taboo against UFOs was fixed, and today, forty years later, that ban on taking UFOs seriously is thoroughly embedded in our society, like an efficiently metastasized cancer.

CHAPTER 12

Taking the Phenomenon Seriously

In order to evaluate the U.S. government's actions and put them in perspective, we can learn a great deal from examining the activities of other governments and their handling of military and aviation UFO encounters. Since the close of Project Blue Book, the United States has become somewhat of a pariah on the international scene when it comes to official UFO investigations, which is especially a problem since as a superpower it has unique potential to influence scientific progress on issues of global significance. Other nations have behaved admirably when UFO events occurred within their airspace. Some have collected useful data when anomalous objects appeared on radar or left marks on the ground, as has happened in France and the UK. These two countries were especially well equipped to handle events as remarkable as a UFO touching down, because they had in place government agencies specifically tasked with taking UFO reports and conducting investigations. Even after the United States bowed out of the UFO business in 1970, other countries kept at it, and still others formed new investigative offices later on, approaching the problem straightforwardly and responsibly.

During the years following the United States' shutdown of its only public UFO agency, those moving forward elsewhere have done the best they could, while sometimes struggling for funding and resources. Thankfully, they have not modeled themselves after Project Blue Book. Rather than devote themselves to disseminating false explanations and other propaganda, these agencies have been willing to conduct honest investigations and acknowledge, particularly in cases documented by pilots, the presence of something unidentified that could not be explained. Pilots and air crews in other nations are not pressured to keep quiet, as their American counterparts were during the O'Hare incident, and are not nearly as wary of ridicule as are their American peers. Elsewhere, military and commercial pilots go on the record about their encounters, and press conferences are held to release information. Aviation safety issues are addressed in connection with UFO events. In general, although the U.S. government hasn't budged since 1970, much of the rest of the world has been moving increasingly in the direction of taking UFOs more seriously.

The UK's study of UFOs began in 1950 within the Ministry of Defence, making it one of the longest running official programs in the world. The MoD had a designated agency, or "UFO desk," that handled UFO reports and investigated cases. In December 2009, the staff became so overwhelmed by the volume of UFO reports from the public, which were at a ten-year high, and the endless stream of FOIA requests about the subject that it closed down its public reporting program. The MoD had not found a way to solve these cases, which, it stated, did not represent a national security threat. It did acknowledge, however, the obvious: that any "legitimate threats"—cases involving military pilots, air defense installations, or objects tracked on radar—would still be dealt with accordingly. The UK had also already begun the lengthy process of releasing all the files accumulated during the years the UFO desk was in operation.

In South America, Chile and Peru set up new government agencies tasked with studying UFO cases in 1997 and 2001, respectively. The Brazilian military has conducted UFO investigations since the late 1940s. Russian cosmonauts, scientists, and high-ranking military officials have spoken publicly about UFO events there. And for the first time, the Mexican Defense Department provided data on an unsolved sighting by an Air Force crew to a civilian researcher in 2004, an important step in government openness within that country.

The French government is generally recognized for maintaining the most productive, scientific, and systematic government investigation of UFOs in the world, continuing without interruption for over thirty years. The agency, now called GEIPAN (Group for the Study and Information on Unidentified Aerospace Phenomena), is part of the French national space agency known as CNES, the French equivalent of our NASA, and serves as a model for other nations that have consulted with it over the years. Particularly remarkable is the network of scientists, police officials, and other specialists that are linked to GEIPAN, ready at a moment's notice to assist with the investigation of any UFO case. Its purpose has always been purely as a research agency, not primarily concerned with defense issues as was the MoD in England or with aviation safety like Chile. It was set up seven years after the close of Project Blue Book, and states its mission as simply to investigate "unidentified aerospace phenomena" and make its findings available to the public.

Jean-Jacques Velasco of France, Nick Pope of the UK, and General Ricardo Bermúdez of Chile have all headed small government agencies within their own countries that worked full-time on investigating UFO cases. They, among others writing in the pages that follow, describe their innovative work on behalf of their governments, and the impact such close-up work with the UFO phenomenon has had on their lives. In countries around the world, witnesses and investigators such as these are very aware of the need for greater participation by the United States, and are now coming together to address that problem.

Whether they have set up specific offices for UFO investigation or not, many governments have accumulated massive amounts of UFO case documentation over the decades and the public has placed great emphasis on gaining the release of these official files.

In recent years, as if part of a trend toward greater transparency, unprecedented numbers of these documents have been declassified and made public for the first time. Since 2004, the governments of Brazil, Chile, France, Mexico, Russia, Uruguay, Peru, Ireland, Australia, Canada, and the United Kingdom have released once-secret files, and in 2009 even Denmark and Sweden joined the trend by releasing over 15,000 files each. However, none of these new records have changed our overall understanding of the phenomenon, beyond confirming that the same events occur around the world and that the behavior of the objects, and often of the governments responding to them, has been repeated over and over. Unfortunately, there has been little forward motion in terms of actually solving the mystery, and the acquisition of even more documents is not the answer.

In fact, government investigators have by and large been limited by the fact that all they've been able to do so far is learn as much as possible after a single event is *over*. Without greater resources, not much can be done except for the filing of reports, year after year. Letters from civilians about isolated, often questionable sightings are also added to the aggregate, making up a large proportion of the released pages. Although often fascinating, government documents no longer reveal anything new, and the thousands and thousands of pages have not led to a major breakthrough in understanding. The most sensitive files—the intelligence reports that are concerned with more serious national security implications and likely deeper investigations and analysis—will not be declassified and released. No long-awaited "smoking gun" document has surfaced.

I believe that a demand for the release of yet more files—even in the United States—is no longer a useful focus. It's an interesting sidetrack, but it does not speak to the heart of the problem. Undue emphasis on seeking further release of documents could even prolong the international stalemate we now face, and give governments a way out through claims that they have done their part by declassifying files or will be doing so in the near future.

Yet the public continues to get very excited about seeing new batches of government documents about UFOs. Most recently, the release of large archives by France in 2007 and the UK in 2008, 2009, and 2010 generated a frenzy of international media coverage in America. So many people logged on to the French website its first day that it crashed. Most interesting was the announcement that about 28 percent of the French cases remain unexplained—approximately the same percentage found by Project Blue Book and the Condon report in 1968.

A featured 2008 piece in the *New York Times* by a staff reporter stationed in the UK selectively focused on a few of the silliest new documents released by the British MoD (letters written *to* the agency by wacky everyday people), and provided readers with the standard ridicule and blatantly biased approach traditionally employed by that noted paper. Ironically, this led to the media breakthrough I had been waiting for: The *New York Times* published the first serious op-ed piece about UFOs in the paper's history. "Unidentified Flying Threats" by former UK Ministry of Defence official Nick Pope offered a rational response to that initial, essentially dishonest story. But once again, none of this publicity changed the political landscape in America regarding UFOs, or did much of anything really, except to make the point that UFOs must be taken seriously.

Unfortunately, we have no way of knowing whether more revealing documents remain closeted away by some governments in secure locations. We know even less about what remains classified in the United States, the most important one of all, and it's highly unlikely that these documents will be provided anytime soon. If a government agency does not wish to release certain sensitive material through the Freedom of Information Act, it won't. So in seeking a new emphasis while attempting to inform and persuade American officials to reevaluate the UFO issue, we can begin by learning from the other countries with established government agencies of their own, and finding out what has been gained from these endeavors. How were these agencies set up, and why? How does their work contrast with that of Project Blue Book? What have they learned about UFOs? What actions have they taken as a result?

First and foremost, we turn to France. Exclusive pieces by General Denis Letty, chair of the COMETA group, and Jean-Jacques Velasco, head of the French government agency for over twenty years, explore these questions. Another noted expert from France, Yves Sillard, is one of the most prominent proponents of cooperative international UFO research in the world. Former director general of the French national space center, CNES, Sillard is currently chairman of the steering committee for GEIPAN. In 1977, while head of CNES, he founded the original French scientific committee charged with the investigation of UFO reports—GEPAN, then with a different name. Sillard has served in many important government and research positions between then and his recent return to GEIPAN. In 1998, NATO appointed him assistant secretary general for scientific and environmental affairs.

In the United States, the National Aeronautics and Space Administration (NASA) is considered in the popular mind to be the country's premier scientific organization with the most knowledge about everything that happens in outer space—a global leader in Earth and space research. CNES has a mandate and an esteem in France that parallel those of NASA here. Responsible for shaping and implementing France's space policy in Europe, CNES, although smaller than NASA, also works on developing space systems and new technologies in cooperation with the European Space Agency, headquartered in Paris. Obviously, the views of the successive directors of either organization—CNES or NASA—are of great significance, whether they deal with the complexities of space exploration or the perplexities of the UFO phenomenon.

Yves Sillard, unknown to most Americans, is a man of stature within the European space community. He founded what has become the world's most effective agency investigating UFOs more than thirty years ago, and still plays a leading role in directing that agency today.

Most important, he has successfully bridged what is usually a gap between scientific space research and UFO investigations, thereby assuring their coexistence within the framework of the French government's national space agency. In 2007, Sillard consolidated his ideas in the landmark book *Phénomènes aérospatiaux non identifiés: Un défi à la science* (Unidentified Aerial Phenomena: A Challenge to Science), written under his direction in collaboration with other scientists. A year later, in 2008, I had the privilege of meeting with him at CNES headquarters in Paris.

Mr. Sillard has provided the following commentary, composed specifically for this volume, summing up the current situation. We must all recognize the power carried by these concise, pointed words, which are highly unusual given the stature of Mr. Sillard in the world community.

> The objective reality of unidentified aerial phenomena, better known to the general public as UFOs, is no longer in doubt. The data recorded by GEIPAN are based on rigorous methods of analysis and control. The aeronautical cases come from competent witnesses, trained to cope with unexpected situations and react calmly.
>
> The climate of suspicion and disinformation, not to mention derision, which still too often surrounds the collection of reports, illustrates a surprising form of intellectual blindness. This is obviously the reason for the silence of many witnesses who do not dare to come forward, and is particularly true for pilots, civilian or military, who fear jeopardizing their careers by speaking out. We must be very open with information, in order to minimize the drama and make it easier for witnesses to file reports.
>
> In addressing UFOs, we must consider the future. One day, through the conquest of space, we will be able to journey outside our solar system, something that is conceivable to us now, through simple extrapolation of our existing technical capacities. For the first time, this potential opens the door to a credible vision of contact between faraway civilizations, considered in the past to be unthinkable.

In spite of some spectacular progress in recent years, today's science will appear very humble when looking back a few centuries from now. The development of science even in the next decades will certainly lead to many new concepts, totally unforeseeable today. What appear to be insurmountable obstacles to more advanced civilizations traveling from exoplanets to Earth will probably appear in a very different light then, and completely new hypotheses, linked to still unborn cosmological theories, will likely have been proposed and realized, completely changing how we view the physical world and the surrounding universe.

Even now—though so far the idea is only hypothetical—what if some unidentified phenomena are discovered to be automatic or inhabited vehicles coming from exoplanets? Shouldn't the famous "precautionary principle" inspire political leaders to at least think about the consequences for every aspect of our society if this hypothesis were to be confirmed? The European Environmental Bureau position is that "the precautionary principle justifies early action in the case of uncertainty and ignorance in order to prevent potential harm." It defines "uncertainty" as "a framework of understanding where we know enough to identify what we don't know." The authors of the COMETA Report initiated the process of offering some commonsense recommendations to the highest civilian and military authorities, in order to prepare them to react in the most appropriate way in case what is today only a hypothesis should tomorrow become a reality. I would recommend greater responsiveness from authorities around the world.

As long as no other credible interpretation has been formulated, let us simply hope that GEIPAN and other agencies can make a modest contribution to this debate and that they will stimulate thinking about these phenomena, the existence of which cannot be contested. And finally, let us hope that our joint efforts will inspire unprejudiced minds to consider the extraterrestrial hypothesis with the seriousness and rigor it deserves, as long as no other credible interpretation has been formulated.

CHAPTER 13

The Birth of COMETA

by Major General Denis Letty (Ret.)

*To learn more about the open approach of the French military to the prob-
lem of UFOs, Major General Denis Letty has provided us with his personal perspective on the historic COMETA Report, explaining why he
felt personally compelled to organize the group's investigation. As I mentioned previously, it was the work of a group of retired French generals
and other officials from that country, coming together to write this report,
that first brought the UFO issue to my attention. General Letty was the
initiator of that effort, a central, driving force behind its completion. In
the report, he and the other authors took the American government to task
for its denials of the existence of UFOs, its harsh treatment of witnesses,
and its excessive secrecy and spreading of "disinformation." They asked
the U.S. government to join France and other countries in a cooperative
venture to investigate the UFO phenomenon, perhaps under the auspices
of the European Union. No response has been forthcoming.*

*Denis Letty, chairman of the COMETA group, is a well-known
former fighter pilot who was head of the French Air Defense, southeast
zone, and the French military mission for the Allied Air Force of Central Europe. A Fifth Wing commander, he also served as Strasbourg
air base commander. In 2008, I was privileged to sit down with General Letty at his home on the outskirts of Paris. He and his wife were
extremely gracious to filmmaker James Fox and me, who, complete with
files, notepads, and a film crew to document our discussions, descended
upon their well-kept, hillside duplex apartment with a stunning view
of the city. Meeting him was a milestone for me personally. Dignified,
gracious, and personable, General Letty was candid and relaxed with
us, yet carried tremendous authority. He's still mystified about the UFO
phenomenon and wants very much to see a resolution.*

As we sat around a table in his living room discussing French cases with the cameras running, Letty addressed the issue of government transparency on UFOs. "I don't think a powerful country like America finds it acceptable to acknowledge that something strange can fly over and the country can't clear the skies of it. Another problem can be panic, created by people imagining that their military can't protect them." I carefully noted his further comments about the U.S. government role: "We are convinced that some governments don't say all they know about the subject, and I mean, of course, the States. That's why we asked for good cooperation from all countries. We're ready to do the research, to work together." The general is convinced that nothing remains hidden within the French military about UFOs, since all the files were released the previous year in order to make that very point. General Letty recently expanded his thoughts for us here.

I first became aware of UFOs in 1965 as a captain in the 3rd staff headquarters of the Tactical Air Force (FATAC) in the city of Metz, when I received all the reports submitted by the national police in the territory of the 1st Area. Some were disconcerting. Since there was no perceptible threat, we simply filed them away. At first I was only a bit taken aback, but then competent pilots I knew personally gradually admitted having been confronted by these phenomena.

One was Hervé Giraud, now a colonel, who in 1977 was flying a Mirage IV with his navigator at about 32,000 feet after dark. They saw an extremely bright light approaching on a collision course, heading straight for them. Giraud radioed military air traffic control, which had no radar track on his scope. He had to bank to the right to avoid the object and then tried to keep in visual contact with it. It moved away, and then either it came back or something identical arrived. Giraud felt as if he was being watched at this point, defenseless, and both men were upset, while the pilot had to maneuver into another tight bank. Still, there was nothing picked up on radar. They returned safely to the base at Luxeuil.

Captain Giraud reported that he perceived that the object was solid and immense, comparing it to running into an eighteen-wheeler at night with all the lights on. It didn't emit any light beams, but glowed with a steady, brilliant white light that obscured any shape behind the illumination.

Two points about this really impressed me. Nothing other than a combat aircraft could perform with the speed and maneuverability of this object. But if it were a combat jet, it would have been registered on radar, especially at that low altitude. In fact, no traffic was picked up by the air traffic controllers anywhere in the area of the Mirage IV. Second, the speed of the object during both encounters was so high during a sharp turn that it would have been supersonic. This means that if it were a combat plane, it would have made a loud sonic boom that would be heard on the ground and in the surrounding area, especially while things were quiet at night. No sound was heard anywhere.

There were other cases involving pilots flying Mirage fighter jets and in-training aircraft. But one more account in particular left its mark on me. In 1979 I learned that Air Force Captain Jean-Pierre Fartek, then a Mirage III pilot, had seen a UFO. It was most unusual, because this was not while he was flying, but had taken place at his home in a village near Dijon, during the day. The object was very low to the ground, at close range. I wanted to meet him to discuss this, and I arranged to do so three months later on the Strasbourg base. On another occasion, I went to his home and visited his wife, as well, who also saw the UFO.

He told me that on December 9, 1979, at around 9:15 a.m., his wife was coming down the stairs to prepare breakfast when she saw a strange disc-shaped object through the window. She called for Fartek to come and look. The object was hovering low to the ground, in front of a row of apple trees, branches of which could be seen behind it; because of that, the captain could measure the distance of about 250 meters (820 feet) from their house. It was approximately 20 meters (65 feet) in diameter and 7 meters thick. The weather was clear, with excellent visibility. I still have the notes that I wrote during the meeting in the presence of Captain and Mrs. Fartek, which say:

- The object looked like two reversed saucers pressed against each other, with a precise contour, a gray metal color on the top and dark blue below, with no lights or portholes.
- It was about three meters from the ground, not stabilized, and then rose to the level of the trees, while continuously oscillating, then went down again slightly and stopped. It went up a little once again, always while oscillating; it tilted and accelerated quickly to

reach a speed much higher than that of a Mirage III, and disappeared.

Captain Fartek and his wife provided many other details. There was a clear delineation between the top and the bottom parts of the craft, and the difference in color could not have been due to effects of the sunlight. The clarity and precision of the shape of the object left no doubt that it was something solid and physical. The disc looked like it was revolving symmetrically around an axis, but the oscillations were slow, as if it were trying to find its balance. It moved without any sound. The witnesses could clearly see the trees towering just behind it, but couldn't tell whether it cast a shadow. Captain Fartek carefully checked for turbulence underneath the object while it hovered, but he couldn't detect any, and it left no trace on the ground. Its departure speed was so extraordinary that it disappeared over the horizon in a few seconds.

Captain Fartek reported this incident to the air guard station at the base. He says that other people also saw the phenomenon but didn't dare report it, such as his neighbors and their children. At the time, the base commander instructed Fartek not to talk about this, because he was concerned about ridicule.

Captain Fartek was very upset by this experience. He told me when we met that the sighting called into question his perception of what were then called "flying saucers," because he had never believed in them. Now, he acknowledged to me, after seeing this craft he could no longer doubt

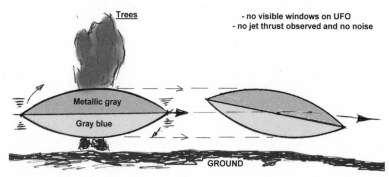

Drawing by Captain Fartek of the object he observed with his wife in 1979. Collection of Jean-Pierre Fartek

their existence. Hearing his testimony, I, too, did not have any more doubt about the reality of the phenomenon. In fact, taken together, I found the Farteks' testimony so disturbing that I have been preoccupied by the UFO problem ever since. In 1996, after he became a major, Captain Fartek was interviewed for the COMETA study that I initiated, and even then, after seventeen years, he was still visibly shaken by what he saw. His case was documented in our report, in the section about sightings from the ground.

The decision to create a twelve-member "Committee for In-Depth Studies," abbreviated to COMETA, to study UFOs, was made in 1996 within the association of veteran auditors of the French Institute of Higher Studies for National Defense, a government-financed strategic planning agency. Since France had been officially studying UFO cases for twenty years, a substantial database of well-investigated and thoroughly documented cases had been gathered by our government agency. In fact, France was a world leader in this process. We felt it was time for an assessment addressing the current situation around the world and defense issues, and the need for international cooperation in dealing with this global problem.

I initiated the private study and became chairman of the group. General Norlain, former commander of the French Tactical Air Force and counselor to the prime minister, and André Lebeau, former head of CNES, were happy to help us and agreed to play major roles. All three of us were retired from the military by this time, although until 2002 I was chairman of an aeronautical company working mainly for French defense.

The investigation lasted from 1996 to 1999. We began by interviewing people who had witnessed UFO phenomena in France and then proceeded to review the best cases that had been recognized and thoroughly studied around the world. We drew on data only from official sources, government authorities, pilots, and the air forces of France and other countries. In the process, we assessed and consolidated the best information and presented our research to the appropriate French authorities.

All the testimony we retained for the COMETA Report is supported by tangible pieces of evidence: radar echoes, tracks on the ground, photographs, electromagnetic phenomena, and even the modification of the process of photosynthesis in plants. Many accounts given by totally independent witnesses confirm one another. It became clear that at least

5 percent of sightings for which there is solid documentation cannot be attributed to man-made or natural sources. Our experts examined all possible explanations for these cases.

We wanted to demonstrate that the UFO phenomenon is real and is not the result of fantasy. I was astonished to discover, and now know for certain, that silent and completely unknown objects sometimes penetrate our airspace with flying capabilities that are impossible to replicate on Earth. And these objects appear to be operated by some kind of intelligence. The COMETA Report shows, in a straightforward manner, that the extraterrestrial hypothesis is the most rational explanation, although of course it has not been proven.

Since the release of the report, I have often quoted General Thouverez, commander of the French air defense force, who in 2002 acknowledged that unknown objects could sometimes be seen in the sky over France and that consequently, it was our responsibility to study them seriously.

Because of statements like this, my co-authors and I believed it was important to submit the COMETA Report to the highest authorities of the state, and we forwarded it to the prime minister and to the military cabinet of the president. In the interests of informing the public, we also published the report in France. At the time of its release, France had reduced the efforts of its national UFO agency at CNES considerably, with only two staff members remaining. After the release of our report, the agency was resurrected and renamed GEIPAN, a process likely facilitated by the support of our group. The COMETA Report has since received worldwide recognition in spite of some virulent denigration by certain people, and when read carefully its findings are impossible to ignore.

We advocated strong international cooperation on UFO investigations, with the United States, in particular, and continue to do so. The sightings in November 2006 above O'Hare Airport near Chicago and over Guernsey in April 2007, which were reported by pilots and air traffic controllers, reinforced our determination not to give up this effort. We now hope that as we continue to collect reports from many colleagues around the world, we will facilitate greater understanding leading to a unified international effort that will determine the true nature and origin of UFOs. We are ready in our country to play a significant role in such an effort.

CHAPTER 14

France and the UFO Question
by Jean-Jacques Velasco

Jean-Jacques Velasco was in charge of the French government's UFO agency for more than twenty years. Although he began his investigations after the close of Project Blue Book, he worked for the French government consistently for about the same length of time as J. Allen Hynek worked for ours. He remained focused and dedicated, as did Hynek, becoming one of the more knowledgeable figures about UFOs in the world. Velasco was an engineer working on the development of French satellites at CNES when he became involved with the new agency studying unidentified aerospace phenomena the year it was founded, 1977, by Yves Sillard. Six years later, he was placed in charge of that agency.

Throughout his tenure, Velasco worked openly within the French national space agency on UFO investigations and was not burdened by a complex, restrictive military framework. He remains actively involved with UFO case studies today and is the author of several books on the subject.

For twenty-one years, from 1983 to 2004, I was the director of the French program to investigate and analyze unidentified aerospace phenomena. Working within the framework of an official mission with specific responsibilities, I had imposed on myself, as was my duty, great reserve in expressing any interpretations or conclusions on the UFO question. Now, all of that has changed. After these many decades of acquired knowledge and experience, I am no longer restricted and can express my personal conclusions with complete freedom of conscience. Therefore, I have chosen to speak here more freely and with more openness than in my previous publications.

First, it is possible to show, using data from established cases officially listed throughout the world, that UFOs—material objects—exist and are distinct from any ordinary phenomena. These cases are few, but their extraordinary characteristics and physical effects demonstrate this fact without ambiguity. On the basis of well-established cases, the existence of UFOs is without question.

UFOs seem to be "artificial and controlled objects," and their physical characteristics can be measured by our detection systems—particularly radar. They display a physics seemingly far different from that which we employ in our most technologically advanced countries. Ground and on-board radar show that their performances greatly exceed our best aeronautical and space capabilities. These capabilities include stationary and silent flights, accelerations and speeds defying the laws of inertia, effects on electronic navigation or transmission systems, and the apparent ability to induce electrical blackouts. When encountered by military aircraft, these objects seem able to anticipate and neutralize pilots' defensive maneuvers, as in such remarkable cases as that of General Parviz Jafari over Tehran and the incidents at Malmstrom Air Force Base. In such encounters, the UFO phenomenon appears to behave as if it is under some kind of intelligent control.

My relationship to this subject matter began in 1977, when I was working as an engineer at CNES, the French space agency. That year, CNES was put in charge of launching an official investigation into the UFO phenomenon in France, under the auspices of a new internal agency then called GEPAN. I soon learned why CNES set up this department—France had been dealing with the question of unidentified aerospace phenomena for more than twenty-five years.

It began in 1951, when three Air Force pilots flying separate Vampire F-5B fighters encountered a shiny, silvery round object. Two tried to close in on it, but it was much faster than they were. A UFO wave followed in 1954, in which gendarmes throughout metropolitan France collected over 100 official reports of "flying saucers," some of which were classified as "close encounters." In one instance, observed by several thousand people, something strange flew back and forth over Tananarive, which today is Antananarivo, the capital of the island of Madagascar. The witnesses were shopping at the outdoor market in the early evening, and were frozen in place and flabbergasted by what they saw. They described a kind of green ball the size of an airplane, followed by a metallic object shaped like a rugby ball.

Dogs were running and howling throughout the city and oxen panicked and destroyed the fences of their enclosures. Most extraordinary was the fact that, during the flight over the capital by this phenomenon, the public power system went off and came back on a few minutes later, after the departure of the "large green ball" and its apparent companion. As might be expected, there was a public outcry and much coverage in the press, all of which prompted an investigation by the French government authorities.

Twenty years later, in 1974, the Defense Minister, Robert Galley, declared on national radio that there existed an unexplained phenomenon that needed to be studied. At the time, I had no idea I would become so involved with this investigation. Our first task at GEPAN, I realized, was to establish a network of police, gendarmerie, Air Force, Navy, meteorologists, and aviation officials and a methodology so that data from sightings could be reported and centralized. A scientific council comprised of astronomers, physicists, legal experts, and other eminent citizens met annually to evaluate and direct studies.

This first phase, from 1977 to 1983, reached three basic conclusions, which still remain valid:

- The vast majority of UFO reports can be explained after rigorous analysis.
- However, some phenomena cannot be explained in terms of conventional physics, psychology, or social psychology.
- It seems highly probable that this small percentage of unidentified aerospace phenomena have a physical basis.

I gradually developed an expertise in these studies, and beginning in 1983 was placed in charge of GEPAN. Following these initial steps, we undertook to develop a more theoretical but still rigorous approach to these studies. It was clear at the outset that it would be necessary to consider both the physical and psychological nature of the phenomenon. In order to fully understand a witness's narrative account, we had to evaluate not only the stated report but also the personality and state-of-mind of the witness, the physical environment in which the event occurred, and the witness's psychosocial environment. GEPAN created a database, unique in the world, of all the cases of sightings of aerospace phenomena recorded by the French authorities since 1951, allowing for statistical analysis.

A classification was adopted that places the UAP (unidentified aerospace phenomena) in four categories:

Type A: The phenomenon is fully and unambiguously identified.

Type B: The nature of the phenomenon has probably been identified but some doubt remains.

Type C: The phenomenon cannot be identified or classified due to insufficient data.

Type D: The phenomenon cannot be explained despite precise witness accounts and good-quality evidence recovered from the scene.

In Type D cases, those which remain unexplained, a subcategorization was also adopted using the "Close Encounters" classification established by Dr. J. Allen Hynek, based on the sighting distance and the effects generated by the phenomenon.

These on-the-spot investigations, carried out at the request of the police or the civil and military aviation authorities, followed by scientific analysis, made it possible to confirm the existence of rare physical phenomena, classified as unexplained UAP, that do not conform to any known natural or artificial phenomena. The statistical analyses and the surveys carried out since the creation of the GEPAN make this even clearer. The Type D category contained more cases during some unusual periods, called "waves," like the wave of 1954, when nearly 40 percent of the cases in the database belong to this last category.

GEPAN initiated several lines of research involving other laboratories and consultants in countries where similar events were occurring, which allowed for comparison with additional files and databases. We worked on developing improved detection systems, such as image analysis of photographs and video footage.

In 1988, GEPAN became a new agency called SEPRA in order to broaden the mission to include the investigation of all reentry phenomena, including debris from satellites, launches, etc. When an unidentified object left traces or any kind of marked effect on the environment that could be recorded and measured by sensors or instruments, we referred to them as UFOs. Among the physical trace ground cases that have been thoroughly investigated, three have stood up to a rigorous analysis and could not be categorized as involving known objects.

In November 1979, a woman called the gendarmes to say a flying saucer had just landed in front of her house. The gendarmes went to the reported landing site immediately, and GEPAN came also with a multi-disciplinary team of investigators. Another witness provided an independent account of an object alighting. The visible trace evidence included a grassy area flattened in a uniform direction, and plant physiology analysis was subsequently carried out by a respected university. Since this was the first time we had collected soil and plant samples from a presumed landing case, rigorous protocols had not yet been established for their analysis, and no significant results were obtained.

However, that changed with the Trans-en-Provence case, one of the best known cases in France. Around 5:00 p.m. on January 8, 1981, electrician Renato Nicolai was building a small water pump shelter in his garden on a sunny afternoon. He heard a low whistling sound coming from above. Upon turning around, he saw an ovoid object in the sky that approached the terrace at the bottom of the garden and landed. The witness moved forward cautiously to observe the strange phenomenon from behind a shed, but, within a minute, the object rose and moved away in the same direction from which it had arrived. It continued to emit a low whistle. As it flew away, Nicolai saw two round protrusions on the underside that he said looked like landing gear. He approached the scene of the apparent landing and noticed circular depressions, separated by a crown, on the ground. The next day, after noticing how upset he was during the night, his wife called the gendarmerie, which came to his home and found two concentric circles on the ground, one 2.2 meters (7.2 feet) in diameter and the other 2.4 meters (7.8 feet) in diameter, with a raised area between them 10 centimeters (about 4 inches) wide. They gathered soil samples of the traces and control samples from outside the area.

The GEPAN investigators went to the site in Trans-en-Provence a month later, collected additional samples of the compacted soil and nearby plants, gathered control samples, and reinterviewed Mr. Nicolai. The physical traces left by "the object" provided the laboratories with much useful information on its nature, its shape, and its mechanical characteristics.

The biochemical analyses carried out on wild alfalfa from the site revealed a major deterioration of the vegetation, apparently caused by powerful electromagnetic fields. Dr. Michel Bounias, from the National Institute of Agronomic Research, showed that the plant degradation was

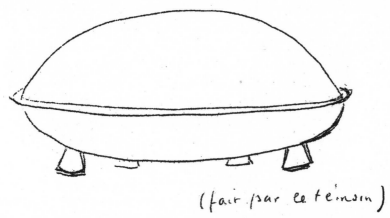

(fair par ee témoin)

Drawing by witness Renato Nicolai of the Trans-en-Provence object, which left visible ground traces and caused nearby plants to degrade. © temoin, GEIPAN

probably due to pulsated microwaves. The following year, new measurements taken on the alfalfa showed a return to normal biological activity.

The GEPAN investigation went on for a full two years and came to some very interesting conclusions. There was evidence of a strong mechanical pressure, probably due to a heavy weight, on the ground surface, and simultaneously or immediately, the soil was heated to between 300 and 600 degrees C. In the immediate vicinity of these ground traces, the chlorophyll content of the wild alfalfa leaves was reduced 30 to 50 percent, inversely proportional to its distance from the landing site. The younger alfalfa leaves experienced the highest loss of chlorophyll, and moreover exhibited "signs of premature senescence." By way of comparison, biochemical analysis showed numerous differences between vegetation samples obtained close to the site and those more distant.

The report concluded that "it was possible to qualitatively show the occurrence of an important event which brought with it deformations of the terrain caused by mass, mechanics, a heating effect, and perhaps certain transformations and deposits of trace minerals." Nuclear irradiation does not seem to account for the observed effects, but some type of electrical energy field might account for the chlorophyll reductions.

Roughly one year after the case of Trans-en-Provence, the so-called Amaranth case of 1982 involved the daytime sighting by a scientist (M.H., a cellular biologist) of a smallish object about one meter in diameter hovering above his garden. The witness first saw the shiny flying craft at 12:35 p.m. in front of his house, making a slow descent. He stepped back as it seemed to move toward him, until it stopped about one meter above the ground and sat there, silently hovering for about twenty minutes, which he measured by looking at his watch. He was not frightened, and, being a scientist, made a detailed, precise observation. He described it as oval and resembling two coupled metal saucers, one on top of the other, the upper half a blue-green dome. It suddenly shot straight up, as if pulled by strong suction, and the grass underneath momentarily stood straight up, but it left no visible traces on the ground.

The gendarmerie made extensive notes on the event within five hours and reported their findings to GEPAN, which sent a team of investigators forty-eight hours later. Of high interest were the visible traces left on nearby vegetation, particularly on an amaranth bush, whose leaves were desiccated and dehydrated after the event. The fruits of other plants around where the object had hovered looked as if they had been cooked. Biochemical analyses showed that these effects could only have been caused by a strong heat flux, most likely due to powerful electromagnetic fields, causing dehydration. This electric field would have had to exceed 200 kV/m at the level of the plant, which could also have caused the blades of grass to lift up. Subsequent investigations showed

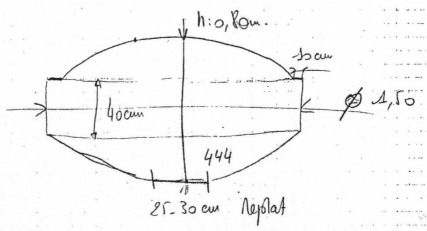

The "Amaranth case" involved an object, drawn by the witness, that hovered near the ground. Vegetation was desiccated, most likely by powerful electromagnetic fields. © temoin, GEIPAN

that this phenomenon could be reproduced in the laboratory by using very intense electric fields.

A psychologist in charge of analyzing the testimony and the psychological profile of the witness concluded in his report that this story had not been invented and that the witness was neither a mythomaniac nor a hoaxer.

Such field investigations demonstrated the possibility of the physical reality of the UAP, but, in fact, the aeronautical cases are the ones which provide the most convincing results on this question. Unlike land witnesses, pilots are operating within the framework of a transportation or air security mission, following the directives coming from civilian or military navigation control centers. They are neutral and highly trained observers when sightings of UAP occur. Such observations of strange unidentified air phenomena by civilian and military pilots in France led to the creation of a database of 150 cases of aeronautical UAP beginning in 1951. The classification into the four categories showed that over 10 percent (fifteen) of the aeronautical UAP cases belong to Type D, the ones that can't be explained despite precise witness accounts and good-quality evidence. In about half these cases, environmental effects such as electromagnetic interference with on-board instruments and/or disturbances of the radio connection with air traffic controllers were reported by the pilots when UAP were nearby.

In January 1994, SEPRA investigated a case that turned out to be the most exceptional pilot case documented in the French skies. On January 28, Captain Jean-Charles Duboc and copilot Valerie Chauffour were piloting Air France flight 3532, making the Nice-London connection at a speed of 350 knots (approximately 650 kilometers/hour) in the early afternoon. The visibility was excellent when a crew member informed the captain and his copilot about a dark object to the left of the aircraft, which he thought was a weather balloon. It was 13:14 GMT and the sun was at the zenith. At first, Duboc thought it was an aircraft banking at a 45-degree angle, but soon all three agreed that this was not a familiar object. They estimated a distance of twenty-two miles (fifty kilometers) at an altitude of six miles (ten kilometers). At first it looked bell-shaped, and then more like a lens or disc, brown and large, and the witnesses were struck by its changes in shape. After about a minute, it disappeared almost

instantaneously, as if suddenly becoming invisible, without any escape trajectory. The duration of this sighting was approximately a minute.

Captain Duboc reported the incident to authorities at the Reims
air navigation control center, which had no information about any aircraft in the location. A report was then sent to SEPRA, which classified
it as Type C, meaning it was insufficiently documented for identification.
However, Reims contacted the Taverny air defense operations center,
CODA, and we later learned something important that allowed us to
reclassify this event as a clear Type D: CODA recorded a radar track at
their control center in Cinq-Mars-la-Pile that corresponded in both location and time to the observation of the crew of Air France flight 3532.
The object disappeared from view of the radar scope and the crew *at the
same instant.* CODA's investigations ruled out the possibility of a weather
balloon. Because the precise crossing distance of the two trajectories was
known, experts estimated that the UAP was about 750 feet long.

In studying aviation cases, an important contribution was made by
an outstanding independent French investigator, Dominique Weinstein, who has catalogued 1,305 cases of UAP and UFO sightings by
pilots—cases for which adequate data is available to categorize the UAP
as unknowns—collected from official sources, including material I provided from CNES/SEPRA. The following results are interesting: 606
cases (36.7 percent) are sightings by military pilots and crews; 444 cases
(26.9 percent) are sightings by civilian pilots; and 196 cases (11.8 percent) are by private pilots. In 200 cases (12.1 percent) the visual observation was confirmed by on-board or ground radar. And in 57 cases (3.45
percent), the pilots noted electromagnetic effects and disturbances on
one or more of the plane's transmission systems.

In combination with radar, we can draw a clear picture of the physicality of the UFO maneuvers in the airspace. The analysis of certain characteristics and maneuvers of these objects indicates behaviors that have
nothing to do with any natural phenomena or with operations carried
out by aircrafts or aeronautical and space machines.

One crucial point I have noted, which is shown in Weinstein's study,
is that a UFO's behavior tends to depend on whether the encounter
involves a military aircraft or a civilian passenger plane. Neutrality usually seems the general rule with commercial airlines or private planes,
whereas an active interaction often occurs between UFOs and military

aircraft. Military pilots usually describe the movements of UFOs as they would air maneuvers of conventional aircraft, using terms such as follows, flees, acute turns, in formation, close collision, and aerial combat. Twenty-two military cases in the Weinstein catalogue involve near misses, and six include reported "dogfights," or combat maneuvers, between the UFOs and the military aircraft. I conclude that these incidents clearly demonstrate that in no way are these examples of natural events, but rather that UFOs are phenomena with a deliberate behavior. The physical nature of UFOs has been proved. Some of them also exhibit *intelligent control* when interacting with military aircraft.

I would like to propose an intriguing hypothesis that is important to me personally. On my part, it has required some research that extends outside of France and into the United States. I believe that there is a connection between strategic nuclear power, the atomic bomb, and the presence of unidentified artificial objects in the sky. This is suggested by data collected over several decades. It could be part of the answer to the question of why UFOs have been present in our environment.

I find it very interesting that this association between the sensitive strategic sites and the overflights of "flying discs" was proposed within the American Air Force during the Cold War. Air Force intelligence noted that many sightings occurred over "sensitive installations." According to one document, a meeting was held on February 16, 1949, in Los Alamos, New Mexico, that included Edward Teller, "the father of the H bomb." Commander Richard Mandelkorn of the U.S. Navy wrote in his report on the meeting that "there is cause for concern of the continued occurrences of unexplainable phenomena of this nature in the vicinity of sensitive installations." And an Army intelligence memo written a month earlier outlining different theories for these "extraordinary phenomena" stated almost the same thing: "It is felt that these incidents are of such great importance, especially as they are occurring in the vicinity of sensitive installations." On April 28, 1949, Dr. Joseph Kaplan, member of the Air Force Scientific Advisory Board, recommended a scientific investigation about the observed "unidentified aerial phenomena" and emphasized that "this was of extreme importance" because "these occurrences relate to the National Defense of the United States."

Such historical documents enable us to understand the origins of the connection between UFOs and nuclear bases, and to see that this problem was taken very seriously by the military and governmental authorities. Most explicit was part of a report by George E. Valley, MIT physicist and radiation expert and member of the Air Force Scientific Advisory Board, submitted to the Air Force Project Sign in 1949. Valley swept aside all the assumptions of known natural and artificial phenomena and advanced the hypothesis of extraterrestrial objects, specifically "space ships." He states that any "extraterrestrial civilization" making these objects would have to be developed far in advance of ours. He goes on to write:

> Such a civilization might observe that on Earth we now have atomic bombs and are fast developing rockets. In view of the past history of mankind, they should be alarmed. We should, therefore, expect at this time above all to behold such visitations.
>
> Since the acts of mankind most easily observed from a distance are A-bomb explosions, we should expect some relation to obtain between the time of A-bomb explosions, the time at which the space ships are seen, and the time required for such ships to arrive from and return to home base.

We have on record the number of explosions worldwide and tests both in the atmosphere until 1963 and underground from 1958 to 1998, from the first explosion in the New Mexico desert in 1945 to the most recent in India in 1998, a total of just over 2,400 explosions (543 atmospheric tests and 1,876 underground explosions). By comparing nuclear tests to some 150 visual/radar UFO cases collected since 1947, we note that the curves are practically superimposed in time and that they coincide, with not more than a few months appearing between the number of explosions and one of the UFO appearances. This similarity in the two curves would suggest that the proven presence of UFOs is related to the nuclear strategic activity in the world. I base my hypothesis on my studies of official documents, the places and zones of UFO sightings, and remarks made by highly placed civilian and military persons involved in secret programs. There have been numerous instances of UFOs flying over or near strategic air command and other military bases in the United States, especially as documented during the 1960s.

In fact, flights of "green fireballs" and "flying discs" occurred over sensitive U.S. sites such as Los Alamos, Albuquerque, Kirtland AFB, Alamogordo, and Holloman AFB. The perimeters of Oak Ridge, Hanford, and Knoxville, where the materials intended for the nuclear bombs were produced, were also flown over. And other examples have been documented: Great Falls and Malmstrom AFB (Montana); Fairchild (Washington); Kincheloe, Wurtsmith, and Sawyer AFB (Michigan); Plattsburg (New York); Loring AFB (Maine); and Pease AFB (New Hampshire). Perhaps if there is some kind of monitoring going on, it manifests more strongly when there is a nuclear crisis situation on the planet. On March 16, 1967, at Malmstrom Air Force Base in Montana, nearly twenty nuclear missiles were suddenly shut down while UFOs were in close proximity.

Something very extraordinary also occurred one year earlier at Minot Air Force Base in North Dakota: On October 24, 1966, the Minuteman missile system was adversely affected during an afternoon while UFOs were sighted from the ground by multiple observers at three separate missile sites for over three hours, and two objects were tracked on radar. Communications and radio transmissions between various facilities monitoring the events were disrupted by static when the UFO came close to the site.

At 4:49 p.m. the outside and interior security alarms of safety for the Oscar 7 missile silo were activated at the control desk located sixteen kilometers (ten miles) away. A security team was dispatched and discovered that not only was the fence open but the horizontal door closing the missile silo was also open. This reinforced-concrete door weighed nearly twenty tons and there were no tire tracks nor any record of a visit that could account for this.

This case puts in stark view some serious questions about the nature of this phenomenon that was responsible for: various ground and on-board radar echoes; the loss of the UHF transmissions; the simultaneous observation on the ground and from the air of this immense stationary luminous ball above the Oscar 7 zone; the alarm trigger; and the rising of the twenty-ton silo door. The main witnesses to this incident were located and interviewed years later, confirming these events. The Minot Air Force Base director of operations submitted a detailed report, released with the Air Force Project Blue Book files.

Unlike the Tehran case in 1976, where the Iranian military authorities did not know how to react in the presence of UFOs, the U.S. Air Force knew that it should not suddenly intervene by force above a Minuteman missile silo, but instead should remain as neutral as possible faced with this kind of situation.

I am fascinated with the possible correlation between nuclear activity, the location of nuclear weapon storage facilities, and the presence of UFOs. We can see on a graph the relationship between atomic explosions and visual/radar sightings, by looking at the similarity in the two curves. We can't be certain why, but perhaps UFOs are "monitoring," and this activity was heightened during times of dangerous nuclear activity on the planet.

After my many years studying the most important unexplained cases, I think we have reached a certain level of knowledge about UFOs. They seem to be artificial and controlled objects whose physical characteristics can be measured by our detection systems, radar in particular. They fall under a physics which is by far superior and more evolved than the one we have in our most technologically advanced countries, highlighted by the stationary and silent flights, the accelerations and speeds defying the laws of inertia, the effects on the electronic navigation or transmission systems of aircraft, and the electrical blackouts. These performances have been shown on radar. When military aircrafts are directly involved, these objects are able to anticipate and neutralize the maneuvers of the pilots assigned to security and defense missions, and some remarkable cases show the capacity of the UFOs to seemingly understand a particular situation or to anticipate intentions of escape or military neutralization. The UFO phenomenon is definitely related to something controlled and intelligent.

The only speculation that I allow myself to make about UFOs is that if they are artificial probes, they cannot be of terrestrial origin and consequently they must come from somewhere else. If extraterrestrial civilizations exist and have the capability to reach us, their motivation might be to monitor our planet because of the concerns raised by human behavior.

CHAPTER 15

UFOs and the National Security Problem

While the French agency under Velasco's direction was focused on the scientific study of UFO evidence as a program within the National Space Center throughout the 1970s, '80s, and '90s, the American government was doing absolutely nothing to address ongoing UFO sightings on the other side of the Atlantic, no matter who reported them or what effect they were having on aircraft or military facilities. Since the termination of Project Blue Book, U.S. public policy seemed to be to deny any interest in UFOs whatsoever, even if it meant obvious evasiveness or a little bending of the truth here and there. Ideally, despite the extraordinary data collected in France and other parts of the globe, the U.S. government clearly hoped that everyone in America would simply forget about UFOs altogether.

Air Force statements issued at the close of Project Blue Book generated ammunition for UFO denial that is still used today, showing that nothing has changed in America for over forty years. When approached with a question about UFOs, the Air Force still sends out essentially the same form letter—ironically called a "fact sheet"—that it began using when Blue Book was terminated. Stating that UFO investigations have been discontinued, the statement presents three points—*exactly* the same ones made by the Air Force in its 1969 news release announcing the close of Blue Book. It stated then, as it does today, that the U.S. government will no longer be investigating UFOs for the following reasons:

- No UFO reported, investigated, and evaluated by the Air Force has ever given any indication of threat to our national security.
- There has been no evidence submitted to or discovered by the Air Force that sightings categorized as "unidentified" represent

technological developments or principles beyond the range of present-day scientific knowledge.
- There has been no evidence indicating the sightings categorized as "unidentified" are extraterrestrial vehicles.

Did this Air Force "fact sheet" really give us the facts at the time, and is it applicable today? In contrast to other government agencies that are represented in this book, a look behind the scenes at how the American government *really* has behaved toward UFOs since the close of Blue Book—despite its public positioning—shows continuing official duplicity and leaves many questions unanswered about what was actually going on.

In examining the fact sheet, the second point can be disputed simply by credible, multiple-witness case studies on record at the time, and many others that have occurred since, such as those of General Parvis Jafari and Comandante Oscar Santa María Huertas. Dr. James Harder, a University of California professor of civil engineering, told the House Science and Astronautics Committee in its 1968 hearing: "On the basis of the data and ordinary rules of evidence, as would be applied in civil or criminal courts, the physical reality of UFOs has been proved beyond a reasonable doubt." UFOs have demonstrated "scientific secrets we do not know ourselves." The question of extraterrestrial origin, the third point, remains an unproven hypothesis, but there *was* enough evidence at the time to keep this possibility in the running, and certainly no justification for dismissing it altogether. The first point, a claim that UFOs have never threatened national security, however, is the one most relevant to any government, because it absolves agencies charged with defending the nation from any responsibility for paying attention to unidentified objects in the sky.

However, this first point is simply false. *No* UFO, not even one, has ever impacted national security? "Threat" may be too strong a word, and it could be that the choice of that particular word, as uttered by General Samford in his 1952 press conference, is what allowed the Air Force to get by with the statement that no UFO has ever given even an *indication* of threat to national security. We still have not observed hostile or aggressive behavior from a UFO. But there is no question that in the years leading up to this statement, UFOs *had* shown themselves to be of defense or national security concern, impacting our defense capabilities and causing alarm during the Cold War.

Despite the Robertson Panel intent to diminish public focus on UFOs for national security reasons, former CIA director Vice Admiral Roscoe Hillenkoetter, the first director of the CIA, who served until 1950, did not agree with the 1953 CIA position that UFOs should be ridiculed in the public arena. In 1960, he issued a statement, as reported in the *New York Times*. "It is time for the truth to be brought out in open Congressional hearings," he said. "Behind the scenes, high-ranking Air Force officers are soberly concerned about the UFOs. But through official secrecy and ridicule, many citizens are led to believe the unknown flying objects are nonsense. To hide the facts, the Air Force has silenced its personnel." The opening of the article, distributed through United Press International, reads as follows:

> The Air Force has sent its commands a warning to treat sightings of unidentified flying objects as "serious business" directly related to the nation's defense, it was learned today. An Air Force spokesman confirmed issuance of the directive after portions of it were made public by a private "flying saucer" group. The new regulations were issued by the Air Force inspector general Dec. 24. The regulations, revising similar ones issued in the past, outlined procedures and said that "investigations and analysis of UFOs are directly related to the Air Force's responsibility for the defense of the United States."

Later that year, Congressman Leonard G. Wolf entered an "urgent warning" from Vice Admiral Hillenkoetter into the Congressional Record, stating that "certain dangers are linked with unidentified flying objects," particularly since UFOs could cause accidental war if mistaken for Soviet weapons. He pointed out that General L. M. Chassin, NATO coordinator of Allied Air Services, warned that a global tragedy might occur. "If we persist in refusing to recognize the existence of the UFOs, we will end up, one fine day, by mistaking them for the guided missiles of an enemy—and the worst will be upon us," he said. Based on a three-year study by the well-known National Investigations Committee on Aerial Phenomena (NICAP) with which Hillenkoetter was associated, Rep. Wolf stated that all defense personnel "should be told that the UFOs are real and should be trained to distinguish them—by their characteristic speeds and maneuvers—from conventional planes and missiles. . . . The American people must be convinced, by documented facts, that the UFOs could not be Soviet machines."

Later, a different type of national security concern was registered that didn't involve the Russians, but concerned the safety of our own military bases. Just two years before the Air Force told the public that UFOs were not a national security threat, an event occurred which some former military officers believe dramatically contradicts that conclusion, even though any *intent*—purposeful or directed action—on the part of the UFO remains impossible to determine.

On the morning of March 24, 1967, Air Force First Lieutenant Robert Salas, a missile launch officer, received a call from a frightened security guard reporting a glowing red, oval-shaped object hovering directly over the Oscar Flight Launch Control Center at Malmstrom Air Force Base in Montana. With an "above Top Secret" clearance, Salas was stationed there as part of a team in charge of the missile sites and responsible for deploying the nuclear-tipped warhead missiles in the event of a war. Salas immediately went to wake up the crew commander, First Lieutenant Fred Meiwald, who was napping on his break. Then, within one minute of the phone call, the missiles started shutting down, one by one.

"They went into no-go while the UFO was overhead," Salas says. "This means they were disabled, not launchable." There were ten missiles at Oscar Flight, and Salas remembers losing all of them. The missiles were located five to ten miles from the control center where the UFO hovered, and were about a mile apart from one another with independent backup power sources. A week earlier, on the morning of March 16, 1967, about thirty-five miles away from Oscar Flight, UFOs had visited the Echo Flight facility as well, and all of its missiles went down, too. In total, twenty missiles were disabled within the span of a week.

A formerly classified Air Force telex states that "all ten missiles in Echo Flight at Malmstrom lost strat alert [strategic alert] within ten seconds of each other. . . . The fact that no apparent reason for the loss of ten missiles can be readily identified is cause for grave concern to this headquarters." Salas learned from Boeing engineers years later that technicians checked every possible cause for the missile failures, but were not able to find any definitive explanation for what happened. At the time, it was suggested that the most likely cause would have been some kind of electromagnetic pulse directly injected into the equipment. Whatever force was involved had to penetrate sixty feet underground to do its damage.

In 1995, when Lieutenant Salas attempted to access government files about the incident, the Air Force sent him its reissue of its 1969 public statement—today's "fact sheet"—that no UFO has ever given any indication of threat to our national security, with a letter stating that this statement still held true. Given his experience, and subsequent confirmation by other witnesses about the 1967 Malmstrom incident, Salas clearly disagrees with this national security assessment. "It is simply incorrect," he says. "If you consider the fact that this UFO incident resulted in the loss of twenty missiles during the Cold War and the Vietnam War, this *was* a national security threat. The Air Force is not telling us the truth." Salas is not the only former Air Force officer to take this position. Others—missile personnel, security police, radar operators, and pilots—have come forward with similar reports.

We can conclude that the Air Force statement justifying the close of Project Blue Book was based on falsehoods about issues of great importance to the American people at the time. The denial of the real picture on UFOs was in itself dangerous. And it doesn't make sense. Could the U.S. military *really* have decided to turn its back on UFOs in 1969, when sightings impacting air bases were occurring? It seems inconceivable. This would have been highly irresponsible, a breach of duty. More likely, our government misinformed the public in order to take UFOs out of public view. The escalating public demands for answers to something that the Air Force could not explain in the late '60s were burdensome, and the CIA's strategy of "training and debunking" had not been quite enough to take care of the problem. Perhaps the authorities in charge wanted to quell fears about any possible hazards associated with UFOs, since they couldn't do much about them anyway. But it seemed highly unlikely that all official UFO investigations were simply dropped.

Now we no longer have to speculate about that question, thanks to an explosive government document, once classified, that was later released through the Freedom of Information Act. Issued secretly two months *before* the 1969 Air Force announcement that all government UFO investigations would be terminated, it shows that, in fact, UFOs *were* considered to be a national security issue and would continue to be treated as such. The October 1969 "Bolender memo," as the document has come to be known, illustrates the duplicity of the government's public stance on UFOs.

The purpose of the memo, as sent by Air Force Brigadier General Carroll H. Bolender, a former World War II night fighter pilot who later became NASA's Apollo mission manager, was to officially terminate Project Blue Book. In doing so, Bolender made the point that regulations were already in place through which "reports of unidentified flying objects which could affect national security" are made, those reports that are "not part of the Blue Book system." This suggests that even before the close of Blue Book, the more sensitive reports were already being channeled elsewhere. It goes on to say that "the defense function could be performed within the framework established for intelligence and surveillance operations without the continuance of a special unit such as Project Blue Book." And further:

> Termination of Project Blue Book would leave no official federal office to receive reports of UFOs. However, as already stated, reports of UFOs which could affect national security would continue to be handled through the standard Air Force procedures designed for this purpose. Presumably, local police departments respond to reports which fall within their responsibilities.

In other words, the military really didn't need Blue Book—simply a public relations operation anyway—to continue dealing with UFOs. Instead it would, without public scrutiny, keep the necessary case investigations going, telling the people that there had never been an indication of a national security threat from any UFO. Three important points are made clear in the Bolender memo, unknown to most Americans and likely most government and military officials at the time, which tell us the real government position:

- UFOs *can* affect national security.
- A "defense function" may be necessary in responding to UFOs.
- Reports affecting national security are "handled" irrespective of Project Blue Book.

We don't know to what extent the low-ranking officers staffing Project Blue Book, or the more important Blue Book scientist Dr. J. Allen Hynek, knew that some UFO reports were filed and investigated elsewhere. Dr. Condon, in preparing for the release of his study from the University of Colorado, believed that he had access to *all* UFO data in the government's files, and that nothing was kept from him. That appears to be a questionable assumption. Although some Blue Book chiefs had high clearance, it's possible that some national security cases never reached their desks.

After Blue Book was closed, we know that the U.S. government continued to have some level of involvement in UFO investigations through a range of agencies. Despite government statements to the contrary, this fact has been revealed in official documents released later through the Freedom of Information Act. Two glaring examples involve the cases from Iran and Peru of attempts to shoot down UFOs, as recounted earlier by General Parviz Jafari and Comandante Oscar Santa María. U.S. government officials were interested in both cases and filed classified reports on them at the time—reports that show they took these cases seriously but wanted to keep that interest secret.

At home around the same time, in 1975, American officials were still dealing with sensitive UFO activity near Air Force bases in the western United States. The U.S. Air Force scrambled military jets over Montana to chase multiple unknowns, as detailed in the official 24th NORAD (North American Air Defense Command) region senior director's log. The November 8, 1975, log reports the arrival of two to seven UFOs—one "large red to orange to yellow object" with small lights on it and another with white and red lights. "Conversation about the UFOs; Advised to go ahead and scramble; but be sure and brief pilots, FAA," the document says. Two F-16s attempted to approach, but as the fighter jets drew closer, the object's lights went out and came back on only when the fighters departed. Eventually, the object increased speed to a "high velocity," shot upward, "and now cannot tell the object from the stars," the NORAD log reports.

This report has interesting similarities to other cases in which the UFO appears to "react" to approaching Air Force jets. Here, according to NORAD, the lights went off when the planes approached within close range, and then the pilots couldn't see the UFO. When they retreated, the lights reappeared. It seems, once again, that some kind of intelligence responded and devised a means of "escape."

The American military reported all of this among themselves, but kept it away from the American people. And there was more. The next day the log records the sighting of an "orange white disc object," resulting in an order for a "mobile security team" to investigate. Two more were seen on November 12; one "appeared to be sending a beam of light to the ground intermittently" and then disappeared.

Unlike the full reports we have on the Iranian, Peruvian, and Belgian aerial pursuits of UFOs by armed fighter jets, the more abbreviated NORAD logs do not reveal the mission of the U.S. Air Force scrambled jets. Would the pilots have fired at the UFOs if they were close enough and in a position to do so? Did they not consider the objects to be a potential threat to national security? What actions on the part of the objects *could* have provoked Air Force aggression? Defense Department reports state that UFOs were pursued by U.S. Air Force fighter planes after the objects hovered over three supersensitive nuclear missile launch sites, also in 1975, according to the *Washington Post*. "A string of the nation's supersensitive nuclear missile launch sites and bomber bases were visited by unidentified low-flying and elusive objects," the *Post* reported. The sightings were recorded on radar over installations in Montana, Michigan, and Maine. The objects hovered, in some cases as low as ten feet off the ground. "In several instances, after base security had been penetrated, the Air Force sent fighter planes and airborne command planes aloft to carry on the unsuccessful pursuit. *The records do not indicate if the fighters fired on the intruders*," the *Post* continues (emphasis added).

And, it says, during these pursuits, the attempts to "detain" the objects were also unsuccessful. Detain? This is peculiar; how would the military detain one? Chances are, the only way to detain such a craft would be to physically disable it, or shoot it down. The *Post* statement suggests that the Air Force may have tried to do just that, but we don't know, and have not yet been able to find out.

We do know a great deal, however, about what happened in 1976 over Tehran, and about the 1980 incident in Peru, partly because of U.S. government interest in both cases, which led to the filing of American reports with intelligence agencies. One can assume that the cases of Jafari and Santa María must have been of particular interest not only because the pilots took military action against the UFOs, but also because they actually *interacted* with them. In both instances, there was an interplay of actions and *response* over an extended period of time, a type of communication, between a vulnerable man in a small plane and an unknown, highly technological flying machine. Neither pilot knew where it was from, or why it was there. But during this lengthy engagement, both were able to observe the objects at very close range.

The national security aspect is obvious—or perhaps global security would be a more apt phrase. In their attempts to shoot down the UFO, neither pilot was successful, but for different reasons. Santa María penetrated the object the first time with a barrage of shells, which had absolutely no effect, but in subsequent attempts the object shot up vertically extremely fast and avoided additional fire. In Jafari's case, missile firing mechanisms were rendered inert several times at the moment he was about to launch his missiles. Both UFOs demonstrated an uncanny, astonishing trait: They repeatedly evaded attack *at the very last moment*, just when the pilots locked on to the target and were ready to fire, as if they somehow "knew," or registered in some way, when the pilots were about to push the button. These last-minute evasions seem too perfectly timed, and were repeated too many times, to be coincidental. Both cases are among the best illustrations on record of some kind of *intelligent control* by a UFO. Despite the distance between them, the objects appeared to be highly cued in to the actions of the aircraft with which they were engaged. And neither UFO retaliated or harmed the jets, despite their aggressive maneuvers. One would surmise that our government would clearly have been interested in such remarkable events, despite claims to the contrary. And it was. The fascinating FOIA documents tell the real story.

The 1976 Iranian incident was a major news event in Tehran, and even American television was on the scene. As General Jafari described earlier, U.S. Air Force officer Lieutenant Colonel Olin Mooy had attended the debriefing the day after the incident. It was Mooy who penned a three-page U.S. government memorandum titled "UFO Sighting," which was classified and distributed as a teletype from the U.S. Defense Intelligence Agency to the secretary of state, the Central Intelligence Agency, the National Security Agency, the White House, and the Air Force, Army, and Navy. This highly unusual report spells out in detail the information presented at the briefing that Jafari attended, including a description of the primary object and the secondary, smaller objects; the loss of on-board instrumentation in conjunction with attempts to fire; and an apparent landing of one object.

Most significant was the incredible DIA evaluation of Mooy's descriptive narrative, written by Air Force Major Colonel Roland Evans on October 12, 1976. It states:

> An outstanding report: this case is a classic which meets all the criteria necessary for a valid study of UFO phenomena.
> - The object was seen by multiple witnesses from different locations (i.e., Shemiran, Mehrebad and the dry lake bed) and viewpoints (both airborne and from the ground).
> - The credibility of many of the witnesses was high (an Air Force General, qualified aircrews and experienced tower operators).
> - Visual sightings were confirmed by radar.
> - Similar electromagnetic effects (EME) were reported by three separate aircraft.
> - There were physiological effects on some crew members (i.e., loss of night vision due to the brightness of the object).
> - An inordinate amount of maneuverability was displayed by the UFOs.

The evaluation indicates that the reliability of the information was "confirmed by other sources" and its value was High (defined as "unique, timely and of major significance"). It was used, or planned for use, as "current intelligence." This intelligence information of high value, of major significance, concerning an outstanding UFO report that justified further study of the phenomenon, was filed as such—even though U.S. government *disinterest* in UFOs, and outright dismissal of sightings, was the public pattern repeated in so many cases in America, and even though it had told the public in 1969 that UFOs were of no concern.

Four years later, our government also filed a report on the Peruvian incident involving Oscar Santa María. A Department of Defense (DoD)/

Joint Chiefs of Staff "info report" was distributed to almost as many agencies as the Iran report. Titled "UFO Sighted in Peru," the June 1980 document was prepared by Colonel Norman H. Runge, who states that his source was an "officer in the Peruvian Air Force who observed the event . . . source has reported reliably in the past." Santa María does not know the name of that officer, was not interviewed by any American, and clearly remembers that no U.S. officials were present during his briefing. "We were very careful about guarding our own sensitive operations and military procedures," he explained in one of our telephone interviews from his home in Peru.

Unfortunately, the DoD report provides the wrong date for the Peruvian encounter: May 9, 1980, rather than April 11. Santa María believes that the information was distorted, and some of the data imprecise, because the report was not filed until two months after the incident. There were apparently delays as the communication made its way through various channels to the Americans.

The document reports that a UFO was observed over the base, and the Air Commander scrambled an SU-22. "The FAP [Peruvian Air Force] tried to intercept and destroy the UFO, but without success," it states. The pilot "intercepted the vehicle and fired upon it at very close range without causing any apparent damage. The pilot tried to make a second pass on the vehicle, but the UFO out-ran the SU-22."

I find it interesting that the term "vehicle" was used consistently and interchangeably with "UFO" throughout this U.S. government document; usually the term "object" is the official choice, leaving wider room for a range of possible explanations. A "vehicle" is something constructed for the purpose of transporting people or things. This one, which remains of unknown origin, was inexplicably unaffected by large shells fired at very close range. Assuming it was a vehicle of unknown origin, as stated, with a capacity that no man-made vehicle has, the concept of "vehicle" then becomes a provocative one, coming from an Air Force colonel. What was it transporting, and why? There seems to have been no problem with an official acknowledgment of the existence of an actual UFO, ten years after the close of Project Blue Book, as long as the document was classified. In this case, a U.S. Air Force colonel acknowledges the existence of an actual UFO—not what one would expect from a government agency that publicly scoffs at such a notion.

For some reason, in recent decades, there seems to be an official preference for looking into cases overseas, rather than those at home. Perhaps there is a particular interest in military cases, involving either the firing on a UFO or a chase by Air Force fighter jets, which drew the authorities to Iran, Peru, and Belgium. Or is it easier for our government to explore cases overseas without being noticed and drawing attention to a UFO event? If it were to do so openly, the Air Force conclusions issued at the time of Project Blue Book, and repeated ever since, would have to be rescinded. Obviously, the consequences of that would be something the Defense Department would prefer to avoid.

Yet while making these once-classified reports, our officials were well aware of the efforts of governments overseas—the host countries within which they probed for information—to properly investigate military UFO sightings. We benefited from their information, but we certainly did not follow their example.

Instead of contributing in any way, U.S. officials seem to enjoy tip-toeing around the world and checking out cases elsewhere, occasionally finding one that "is a classic which meets all the criteria necessary for a valid study of UFO phenomena," as the DIA evaluation stated. Rather than assigning the U.S. Air Force to openly handle UFO events here, our government has stubbornly ignored unambiguous sightings affecting the lives of thousands of Americans. Simultaneously, it has in place a reporting system for UFO incidents "within the framework established for intelligence and surveillance operations," which it doesn't like to talk about. It all gets a bit confusing. But as far as all of us citizens are concerned, government agencies still provide untenable explanations for American UFO events, or ignore them altogether, even when close encounters raise issues of aviation safety and, yes, national security, and even though we know they are interested in cases overseas. For how long will the authorities continue to toss out the faulty Air Force "fact sheet" to justify this irresponsible behavior?

CHAPTER 16

"A Powerful Desire to Do Nothing"

Most Americans are not aware that, not too long ago, while our government was quietly filing reports on cases overseas, a dramatic UFO wave unfolded over American soil. The spectacle of this wave was as dramatic as the one in Belgium, and the large, low-flying craft resembled those seen over that country in some ways. Only three years after details about the 1980 incident in Peru were distributed to U.S. government agencies, the "Hudson Valley wave" began in upstate New York and parts of Connecticut. It lasted a few years, and after it all died down, our government filed another secret document about the 1990 Belgian wave. But no officials made inquiries about something *here* that happened in between these two other reported events, even though our own UFOs were witnessed by thousands of American citizens. No official documents have been filed about the Hudson Valley events—at least as far as we know.

Yet its resemblance to the Belgian wave was notable. Beginning in December 1982, the American wave also lasted many years, with its peak occurring within a two-year period, and it, too, involved repeated visits by large silent objects, sometimes more than one at a time, hovering at low altitudes with extremely bright spotlights. Groups of people watched, often at close range or while standing directly underneath, and some reported seeing a dark, solid structure behind the lights. Many, while driving along the Taconic Parkway or meandering alone down curving backcountry roads, pulled over to get a better look at the UFOs, while others saw the objects when walking their dogs or jogging along reservoirs and lakes. Witnesses said these structures appeared to be as huge as football fields and were capable of shooting off at incredible speeds from stationary positions. As is typical for UFOs, they were silent or emitted a low humming sound.

The Hudson Valley UFOs, like those in Belgium, did not exhibit any aggressive or hostile behavior. In fact, in similar fashion, the less intimidated witnesses reported flashing their car lights at the objects and receiving flashes in return. And this wave, too, featured simultaneous sightings by police officers—in Danbury, Connecticut, police initially joked about calls from witnesses, before being rudely awakened, which was exactly how the Belgian police initially responded. Later, twelve officers from this department alone had their own sightings. Pathways could be determined due to the volume of reports from varying locations within short time

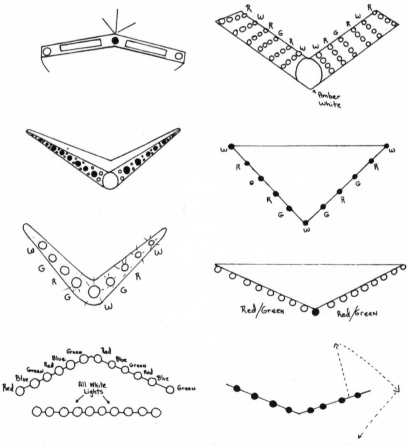

An artist's copy of drawings made by independent witnesses in different parts of New York and Connecticut in 1983 and 1984, which have been reduced and sized to fit the same scale. Collection of Phil Imbrogno

periods, and route maps were constructed just as they would later be in Belgium. Similarly, some nighttime photos and videos were taken in New York and analyzed by various laboratories, although not as extensively, nor were the images as powerful, as the Petit-Rechain photo of 1990.

Although the Hudson Valley residents reported mainly delta- or V- shaped objects and the Belgians saw mostly triangular ones, in reading the many witness accounts of both events, the similar behaviors of the crafts are striking. The bizarre and highly unusual "red light ball" phenomenon reported by the four Belgian policemen made an appearance in upstate New York as well. During the first, dramatic night of the Belgian wave in 1989, two pairs of policemen in different locations watched the red light ball shoot out on a beam from a hovering craft, which was then drawn back into the UFO—a rare detail observed at very close range. Heinrich Nicoll, one of the policemen who witnessed this spectacle, interpreted it to be a probe of some sort. In an interview, he said, "The ball kept leaving and coming back, as if the ball were trying to measure something."

During the Hudson Valley wave, David Athens, chief of the New Fairfield Fire Department in Connecticut, was standing outside talking with a police officer in July 1984 when both saw a row of lights in a circular pattern. "I would say it was something man-made except that two of the red lights dropped down from the group and went in a different direction behind the mountains. One came back and the other didn't," Athens reported.

Jim Cooke, a biomedical engineer, was shocked to see a triangular object hovering no more than fifteen feet above the water of the Croton Falls Reservoir late one October night in 1983 while driving home. He got out of his car and watched from the edge of the water. "Something came from the underside of the object, a red beam of light or something solid that was glowing red—I really don't know what. But it seemed to be probing the water," he said. According to Cooke, the object moved slowly over the reservoir, and at each stop the "red probe" interacted with the water, and was then retracted. Like the Belgian craft displaying essentially the same thing, this one was triangular. Heinrich Nicoll's description was remarkably similar to Cooke's. He also witnessed the phenomenon over a body of water, which he, too, interpreted to be a probe of some sort. We may never know the purpose of this strange red offshoot of the UFO, but this suggests that very similar objects may have visited both locations in the 1980s.

Despite the intriguing similarities, there was a major difference between these events in upstate New York and those in Belgium—not in the details of what actually happened, but in the way these extraordinary close encounters, repeated year after year, were handled by the authorities—those in charge of protecting citizens and monitoring unregistered air incursions over populated areas.

We must remember that the 1989–90 UFO wave in Belgium was handled rationally, openly, and responsibly by the government. The Belgian Air Force was called into action immediately, and other agencies, such as the Gendarmerie Nationale (a combination of police and army) and the Belgian equivalent of our FAA, also cooperated in the mobilization to identify the objects. The Air Force was not only responsive, but was even proactive in its investigation, looking for craft on multiple radar systems, scrambling F-16s to intercept one on three occasions, and then holding a press conference to explain all this to the public. In addition, state-of-the-art analysis was provided by a number of laboratories on the superior photograph of a craft, one of the best UFO pictures on record. And to take it one step further, the Belgian Air Force made all its data and every resource, including radar stations and even aircraft, available to a highly competent group of civilian scientists who organized data, interviewed witnesses, and kept extensive records. All of these important developments were covered in the European media, with some reporting in the United States, as well. Through it all, the Belgian government did not hide information, issue false explanations, or ridicule witnesses. In fact, we know that Colonel Wilfried De Brouwer, head of the Air Force investigation, told the people the truth. Much was learned, except for the most important thing of all: the origin and purpose of the crafts themselves.

However, in the United States, our UFO wave wasn't handled at all. Not a thing was done by any branch of our government. There was no national or statewide mobilization. No Air Force F-16s were launched (at least not as a matter of public record). No attempts were made to capture the objects on radar. Nor was there any established partnership with a leading U.S. research organization to collect reports, though such qualified scientific groups were ready and waiting. No government labs analyzed the photographs. No government body convened a press conference to provide Air Force data for a public eager for information. The local media gave plenty of coverage in places where the events were actually happening and were a fact of life, but because no officials were engaged other than local policemen, national coverage was minimal.

When pressed by concerned callers, the FAA told witnesses that they had seen something other than what they saw—recognizable things that made a lot of noise, such as airplanes in formation, or helicopters. Numerous factors rendered this explanation untenable, the most obvious being that sometimes the craft hovered or moved more slowly than planes could fly, often at very low altitudes, and it was usually silent. Hovering helicopters or a group of planes flying in formation are notoriously loud. Also, the UFO was seen on many occasions when there were no planes or blimps aloft, as confirmed by the nearby airport. Sometimes, witnesses saw a massive, solid structure around the lights blocking out the sky behind it, easily distinguishable from conventional aircraft. In 1984, for example, six security guards at the Indian Point nuclear power plant witnessed the UFO hovering about 300 feet over the reactor in restricted airspace. Two guards told investigators it was a solid object bigger than a football field.

Yet U.S. government indifference never changed, despite the fact that what many called the "Westchester County boomerangs" hovered or cruised off and on for years over the Hudson Valley and parts of Connecticut, arrayed with colored lights that sometimes blinked on and off when approaching people. Witnesses were left to handle these events on their own, encounters that were disturbing to some, frightening to others, and awe-inspiring to almost everyone; but no official guidance was offered as to what to do. Police departments in New York and Connecticut were flooded with calls, but how were the small units to respond? They were simply not prepared or equipped to handle something like this, beyond making records of these witness accounts, some from their own officers. Traffic jams occurred on Route 84, a major thoroughfare, as drivers stared at the sky. And the local airports simply told callers that they had nothing on radar and could not confirm the sightings. Commu-

nities were left unassisted in trying to make sense of these absolutely stag-
gering events, and most of the U.S. public never heard anything about
them.

How could something as momentous as these Hudson Valley sightings, repeated year after year, be ignored by our government and swept
under the rug? This indifference is so stunning that one could justify
questioning whether these events actually took place at all. Many would
ask, how could this *really* have happened if I never heard anything about
it? And why didn't I hear about the wave in Belgium for that matter, or
other very credible UFO sightings, if, in fact, thousands of witnesses were
involved? This puzzling situation, prompting legitimate questions about
whether UFOs actually exist, represents one of the primary reasons intelligent, well-informed Americans don't "believe in" UFOs. And for good
reason. A rational conclusion would be that if this were really happening,
we would all know about it.

If the Air Force Project Blue Book were still in effect at the time of
these sightings in New York State, they *would* have been officially investigated, even if not at the level many of us would have liked. It would
have been harder for the Air Force to offer quick, dubious explanations
for these events, which happened repeatedly and at very close range.
Fortunately, the key scientist with Blue Book throughout its twenty years
was still actively investigating UFO cases in the mid-1980s, and was paying attention to the sightings in upstate New York. Although no longer
formally associated with the U.S. government, Dr. J. Allen Hynek began
investigating the Hudson Valley wave in 1984. By that time he was widely
regarded as the world's foremost authority on UFOs as well as an eloquent spokesman on the subject to the American public. These sightings
were the final focus of Dr. Hynek's life—he died in 1986—and he poured
a great deal of energy into confronting the shocking indifference of U.S.
government officials in the face of the repeated, well-documented visits
by some kind of phenomenon. Government apathy, he realized, is what
had kept the story from exploding into the national media.

Despite the fact that he had been at the forefront of many UFO
investigations for more than three decades, the unrelenting Hudson Valley wave seemed to both awe and baffle Hynek beyond anything else.
Nothing quite like this had happened before in America. In a 1985 essay,

he described "hundreds of largely professional, affluent people in suburban areas," whose statements he and others recorded on cassette tapes, as "astonished, awestruck and often frightened" by the bizarre sightings. When flying over the Taconic Parkway, or cruising low over streets and houses, an "utterly strange and possibly menacing object" constituted a serious hazard that should have concerned the FAA, he wrote. For scientists, these events should have been of breathtaking scientific concern, and the police and the media were completely derelict in their apathy and indifference, keeping the whole thing out of public awareness.

To understand how such things could occur without our knowing about them, we need to examine the total inaction by those in positions of responsibility. "It was as if a malady plunged all who encountered it, except the witnesses, into a deadly stupor," Hynek mused. "In the story of the Boomerang sightings, the FAA, the media, the scientists, the politicians and the military all may momentarily have touched the mystery, but it appears that then apathy intervened, sapping all incentive, and left in its place a powerful desire to do nothing."

Like so many today, Hynek wanted to know how and why this shocking inaction occurred. He had been a committed skeptic about UFOs when hired by the Air Force, and with his colleagues in the scientific world had often made fun of people who reported seeing them. Although he initially set out to show there was nothing to any of this "nonsense," he underwent a gradual transformation during his long tenure working for the government. While investigating hundreds of UFO cases and interviewing countless credible witnesses, he came to recognize that there *was* a real, physical phenomenon involved, and a very mysterious one. He described it this way in 1977:

> I had started out as an outright "debunker," taking great joy in cracking what seemed at first to be puzzling cases. I was the arch enemy of those "flying saucer groups and enthusiasts" who very dearly wanted UFOs to be interplanetary. My own knowledge of those groups came almost entirely from what I heard from Blue Book personnel: they were all "crackpots and visionaries."
>
> My transformation was gradual but by the late sixties it was complete. Today I would not spend one further moment on the

subject of UFOs if I didn't seriously feel that the UFO phenomenon is real and that efforts to investigate and understand it, and
eventually to solve it, could have a profound effect—perhaps
even be the springboard to mankind's outlook on the universe.

In 1985, the dedicated investigator was confronting an extreme manifestation of a peculiarly American phenomenon known as the UFO
taboo—the automatic, deeply ingrained refusal to acknowledge that
something so contradictory to what we consider "normal," and therefore
unacceptable to our worldview, could possibly exist no matter what the
evidence shows. In this case, Hynek observed that the taboo is so powerful that it can thwart the duties of groups of otherwise highly responsible
people in positions of authority. He struggled to find some kind of core
answer to this dilemma.

Hynek noted that seeing the otherworldly Westchester County boomerangs caused stress, trauma, and fear among the witnesses. They were
given no answers and felt unprotected by their government, and many
did not want to "go public" about these events for fear of being ridiculed. Rooted in the minds of most people, such as the policemen who
received reports from witnesses and had not seen anything themselves,
was the collective belief that this type of event cannot possibly happen.
The only way out was to label the witnesses "crackpots." And yet thousands of people actually saw the objects. They were faced with the conun-
drum that they *knew* that these events *did* happen, as did others from the
area personally acquainted with witnesses or informed about sightings
from trusted sources, such as local newspapers. Could all of these people
be lying or confused? Or could it be that there was something larger,
more deeply rooted, that kept government officials from truly listening
to these accounts, accepting them as true, and investigating accordingly?

Hynek postulated that, in its inability to accept something as revolutionary as the existence of these inconceivable crafts, our psyche simply
shuts the whole thing out. The impossible reality "overheats the human
mental circuits and blows the fuses in a protective mechanism for the
mind. . . . When a collective breaking point is reached, the mind must
openly disregard the patent evidence of the senses. It can no longer
encompass such evidence within its normal borders." He concluded that,
due to the totally bizarre, shocking, and even traumatic nature of such

an event, there is no energy for action, as if everyone was operating on a dead battery. This dynamic can affect groups of people as a whole, and those in charge were not exempt from its numbing effects. "With apathy goes the ability to accept even the most inane explanations—anything whatever—to stave off the necessity to think about the unthinkable," Hynek wrote.

This may not provide a complete answer, but it touches on the profound nature of the UFO taboo, which manages to keep us in the dark even about events in our own backyard. This primarily psychological phenomenon, set in motion by the Robertson Panel in the 1950s, operates here with much greater strength and tenacity than it does in other countries. It infused the improper management of our Air Force agency, Project Blue Book, until its eventual demise. Then the taboo became integrated and accepted, affecting all levels of government. It's still hard to believe that the Hudson Valley events slipped by, unnoticed by most of us—but in fact, that's what happened. Of course, if our government had responded the same way the Belgian government did when that country was hit by a similar wave, everything would be different. And even more important, if we had set up an agency similar to the one in France, devoted to research for its own sake, even greater knowledge could have been acquired. The UK, our closest ally, had an office in place to receive UFO reports during the time of the Hudson Valley wave, and would have investigated. The U.S. government, albeit responsible for an enormous territory of land and sky in comparison to France, Belgium, or the UK, appears to be operating at one extreme in its ability to turn a blind eye to UFOs.

CHAPTER 17

The Real X-Files
by Nick Pope

The British Ministry of Defence set up its office for UFO investigations in the 1950s, at around the same time that the United States established Project Blue Book. However, the British kept their investigation going much, much longer. Nick Pope was the man assigned to head this government UFO project from 1991 to 1994. His perspective on the phenomenon changed radically during his years of intensive focus on investigations and access to "inside" government information on UFOs. Like the other contributors to this book, he would like to see more involvement by U.S. officials and intelligence agencies.

Pope has become one of the most active former government officials to speak about this issue, sought by media from around the world as a leading expert. He combines a keen analytical mind with a strong interest in the UFO phenomenon, both of which are leavened with a dry, uniquely British wit. He is yet another example of the many officials and military officers who, as they became acquainted with UFO investigations virtually by accident, flexed their skeptical muscles only to find themselves absorbed by the unexpected power of the evidence they had initially expected to disprove. Nick Pope had access to classified files and other highly sensitive information that he is not at liberty to share, which makes his insights and convictions even more intriguing. Still involved with the subject on a semiofficial basis, he recently worked with the British National Archives as a consultant for the ongoing program to declassify and release the MoD's archive of UFO files.

I worked for the Ministry of Defence for twenty-one years, beginning in 1985. At the time, the policy was to move people every two or three years—either on level transfer or promotion—so that every-

body gained experience in a wide range of different jobs: policy, operations, personnel, finance, etc. I'd completed two or three different jobs,
and by the early 1990s, I was working in a division called Secretariat (Air
Staff) and had been seconded into the Air Force Operations Room in
the Joint Operations Centre. I worked there in the run-up to the first
Gulf War, during the war itself, and in the aftermath of the conflict, as a
briefer, preparing material for the key daily briefings to ministers and the
service chiefs. My job was to collect raw data about Royal Air Force (RAF)
operations, and pick out the key things that senior personnel needed to
know: details of any casualties and losses, targets attacked, battle damage assessment, etc. It was while working there that I was approached
in 1991 and asked whether, after I was released from duties in the Joint
Operations Centre, I would like to work on UFO investigations—a post
embedded in another part of the division. I accepted the invitation even
though I knew little about the subject and I certainly had no belief in
UFOs. So while I was open-minded in all my investigations, my start point
was broadly skeptical.

The MoD had been looking at the UFO phenomenon since the early
fifties and has received over 12,000 sighting reports to date. In all that
time, the objectives haven't really changed much. Back in 1950, the MoD
set up the secret Flying Saucer Working Party, composed of specialists
within scientific and technical intelligence, to investigate and assess the
numerous UFO sightings being reported in the media. In 1951 the group
recommended that investigations should be terminated "unless and until
some material evidence becomes available." But that policy was reversed
a few years later following a series of high-profile UFO sightings involving
the military. Two Air Ministry divisions—S6, a civilian secretariat division
on the Air Staff, and DDI (Tech), a technical intelligence division—then
became actively involved in investigating UFO sightings. Their brief was
to research and investigate the UFO phenomenon, looking for evidence
of any threat to the UK.

That policy was still in place when I came on board in the 1990s.
UFO sightings were to be investigated to see whether there was evidence
of anything of any defense significance, any threat to the defense of the
UK, or information that may be of use to us, scientifically or militarily.
Having a UFO project in no way implies a governmental belief in extra-
terrestrial visitation. It simply reflects the fact that we keep a watchful

eye on our airspace and want to know about anything operating in the United Kingdom's Air Defence Region. Many other countries have similar research efforts.

I had access to all the previous UFO files, some of which had been very highly classified, so I had a vast archive of data to assess. This enabled me to undertake various research projects, looking for trends, etc. But the bread and butter of the job was investigating the new sightings that were reported on a virtually daily basis. We used to receive 200 to 300 reports each year.

The methodology of an investigation is fairly standard. First, you interview the witness to obtain as much information as possible about the sighting: date, time, and location of the sighting, description of the object, its speed, its height, etc. Then you attempt to correlate the sighting with known aerial activity such as civil flights, military exercises, or weather balloon launches. We could check with the Royal Greenwich Observatory to see if astronomical phenomena such as meteors or fireballs might explain what was seen. We could check to see whether any UFOs seen visually had been tracked on radar. If we had a photograph or video, we could get various MoD specialists to enhance and analyze the imagery. We could also liaise with staff at the Ballistic Missile Early Warning System at RAF Fylingdales, where they have space-tracking radar. Finally, on various scientific and technical issues, we could liaise with the Defence Intelligence Staff, though this is an area that I can't discuss in any detail.

After investigation, around 80 percent of UFO sightings could be explained as misidentifications of something ordinary, such as aircraft lights, satellites, airships, weather balloons, or planets. In around 15 percent of cases there was insufficient information to draw any firm conclusions. The remaining approximately 5 percent of sightings seemed to defy conventional explanation. The sorts of cases that got into this latter category included UFO incidents where there were multiple witnesses, or where the witnesses were trained observers such as police officers or military personnel; sightings from civil or military pilots; sightings backed up by photographic or video evidence, where technical analysis found no signs of fakery; sightings tracked on radar and sightings involving structured craft seemingly capable of speeds and maneuvers way ahead of even the most advanced aircraft.

Generally speaking, because my terms of reference limited my investigations to sightings in the United Kingdom Air Defence Region, I did not liaise with other nations on this issue. However, on occasion we raised questions about the phenomenon in general or about specific sightings with other countries, through the respective British embassy. I have also met officials from other countries in a private capacity who have been involved in government work on this subject, such as Jacques Patenet from the French CNES GEIPAN unit, and Colonel Aldo Olivero from the Italian Air Force. In the course of these discussions it was clear that our terms of reference and methodologies were broadly similar, as were our conclusions.

The Cosford Incident

On March 30 and 31, 1993, there was a series of UFO sightings in the UK involving over a hundred witnesses, many of them police officers and military personnel. The UFO also flew directly over two Air Force bases. What follows is the extraordinary story of what has been dubbed the Cosford incident.

The first sighting took place on March 30 at around 8.30 p.m. in Somerset. This was followed by a sighting at 9:00 p.m. in the Quantock Hills. The witness was a police officer who, together with a group of scouts, had seen a craft that he described as looking "like two Concordes flying side by side and joined together." The reports came in thick and fast, and when I arrived at work the following morning I received a steady stream of them. It was soon clear that I had a major UFO event on my hands.

One of the most interesting reports came from a civilian in Rugeley, Staffordshire, who reported a UFO that he estimated as being 200 meters in diameter. He and other family members told me how they had chased the object in their car and got extremely close to it, believing it had landed in a nearby field. When they got there a few seconds later, there was nothing to be seen. Many of the descriptions related to a triangular-shaped craft or to lights perceived as being on the underside of such a craft. Indeed, in an apparent coincidence, these sightings occurred three years to the very day after the famous wave of sightings in Belgium that led to F-16 fighters being scrambled to intercept a UFO tracked on radar.

The UK UFO was seen by a patrol of Air Force police based at the

RAF station in Cosford, 150 miles northwest of London. Their official police report (classified Police In Confidence) stated that the UFO passed over the base "at great velocity . . . at an altitude of approximately 1,000 feet." It described two white lights with a faint red glow at the rear, with no engine noise being heard. The Air Force police report also contained details of a number of civilian UFO sightings they had received in the course of making enquiries with other military bases, civil airports, and local police.

Later that night, the meteorological officer at RAF Shawbury—the base that provides advanced training for helicopter aircrew, air traffic controllers, and flight operations personnel for all three of the UK's armed services—saw the UFO. He described to me how it had moved slowly across the countryside toward the base, at a speed of no more than 30 or 40 mph. He saw the craft fire a narrow beam of light (like a laser) at the ground and saw the light sweeping backward and forward across the field beyond the perimeter fence, as if it were looking for something. He heard an unpleasant low-frequency humming sound coming from the object and said he could feel as well as hear this—rather like standing in front of a bass speaker. He estimated the size of the craft to be midway between a C-130 Hercules transport aircraft and a Boeing 747. Then he told me that the light beam had retracted in an unnatural way and that the craft had suddenly accelerated away to the horizon many times faster than a military aircraft. Here was an experienced Royal Air Force officer who regularly saw aircraft and helicopters, telling me about something he said was quite unlike anything he's ever seen in his life. The MoD party line about UFOs being of "no defense significance" was looking decidedly shaky. What was I supposed to say to him, I wondered—"Don't worry, it was probably just a weather balloon"?

For a number of reasons UFOs are notoriously underreported. The two main factors here are fear of disbelief and/or ridicule, and the fact that many people do not know who to contact with details of their sightings. While there were standing instructions that UFO reports sent to military bases, civil airports, and police stations should be forwarded to the MoD for investigation, this national reporting system did not always work. The case file on the March 30–31, 1993, UFO incident makes it clear that there were many more sightings than ever reached the department. One throwaway line from a police report of a sighting in Liskeard,

Cornwall, stated that the object was "seen by other police officers throughout Devon and Cornwall." We can only guess at the number of sightings that went unreported that night.

Because of the similarity between these reports and those repeatedly filed in Belgium in 1989 and 1990, I asked the Defence Intelligence Staff to make some discreet enquiries to the Belgian authorities through the British embassy in Brussels. As I recall, our air attaché was able to speak to General De Brouwer and the two F-16 pilots. It was clear that De Brouwer had done an excellent investigation under very difficult circumstances.

Like De Brouwer, I launched a detailed investigation into the Cosford sightings, the main difference being that the Cosford incident was not a "wave" but a one-time event, as are most UFO cases. I worked closely with the RAF, colleagues in the Defence Intelligence Staff, and personnel at the Ballistic Missile Early Warning System at RAF Fylingdales. One of the first things that I did was order that radar tapes be impounded and sent to me at MoD Main Building in Whitehall. The radar data was downloaded onto standard VHS video cassettes and arrived shortly thereafter. I watched it with the relevant RAF specialists, who told me that there were a few odd radar returns, but that they were inconclusive. Radar is not an exact science, and in certain circumstances, false returns can be generated. Later, a more formal assessment of the radar data was made. Unfortunately, one of the radar heads was not working on primary radar during the reporting period, so only aircraft working secondary surveillance radar could be seen. But with this and with other checks, we were able to build up a picture of all aircraft and helicopter activity over the UK, so that we could factor them into the investigation and eliminate them from our enquiries if appropriate.

The Ballistic Missile Early Warning System at RAF Fylingdales, with its powerful space-tracking radars, was an important part of my UFO investigation. The authorities there quickly alerted me to the fact that there had been a reentry into the Earth's atmosphere of a Russian rocket carrying a communications satellite, Cosmos 2238. We postulated that this was a possible explanation for a cluster of UFO sightings that occurred at around 1:10 a.m. on March 31.

A theory often put forward to explain some of the most spectacular UFO sightings is that they might be prototype aircraft, drones, or other unmanned aerial vehicles. Of course, at any time we will be test-flying various things that you won't see at the big air shows for several years, but the bottom line is that these tests occur in specific areas, so at least within government we can differentiate between black projects—a classified project that is not publicly acknowledged, such as the F-117 stealth fighter program, prior to 1988—and UFOs.

Even so, there had been controversy brewing about the American Aurora, an alleged hypersonic replacement for the SR-71 Blackbird that some journalists and aviation enthusiasts alleged was being flown in British airspace without the knowledge of the UK authorities. So we raised the issue of the March 1993 UFO sightings with the U.S. authorities, through the British embassy in Washington. Was it possible that something had gone wrong with the normal processes for overflight of another country and could our UFO sightings be attributable to some U.S. prototype? The answer I got back—via our air attaché at the UK embassy in Washington—was extraordinary: The Americans had been having their own sightings of these large, triangular-shaped UFOs and wanted to know if the RAF might have such a craft, perhaps as part of a "black" program, capable of moving from a virtual hover to speeds of several thousand mph in an instant. We wish we had! The interesting thing about this was that somebody in the United States was still clearly taking an interest in UFOs, despite the apparent disengagement from the subject in 1969 with the closing of Project Blue Book.

Given the MoD's "no defense significance" conclusion on UFOs, it seems fitting to conclude this section with quotes from MoD documents which contradict the usual stance. In a briefing that I prepared for the division head on April 16, 1993, after the Cosford investigation, I wrote: "It seems that an unidentified object of unknown origin was operating in the UK Air Defence Region without being detected on radar; this would appear to be of considerable defense significance, and I recommend that we investigate further, within MoD or with the U.S. authorities."

My division head was normally skeptical about the UFO phenomenon, but on this occasion he agreed with my conclusion. His April 22, 1993, brief to the Assistant Chief of the Air Staff (one of the UK's most senior RAF officers) stated: "In summary, there would seem to be some evidence on this occasion that an unidentified object (or objects) of unknown origin was operating over the UK."

This is about as close as the MoD will ever get to saying that there's more to UFOs than misidentifications or hoaxes.

The Rendlesham Forest Incident:
A Cold Case Review

Britian's most spectacular UFO incident occurred late on Christmas night 1980 and in the early hours of Boxing Day, when strange lights were seen in Rendlesham Forest, near Ipswich. The many witnesses were mainly United States Air Force personnel based at the joint U.S. Air Force/NATO twin bases RAF Bentwaters and RAF Woodbridge in Suffolk. Even though the events took place on British soil, these bases were U.S. Air Force facilities at that time. Rendlesham Forest lies between the twin bases, and as the Cold War was still decidedly frosty, a UFO sighting at two of the nation's most sensitive military sites was most decidedly of interest.

At the UFO project, I had access to the large MoD file on this incident, which at that time had *not* been released to the public. Even the most basic information on this case was extraordinary, and I decided to launch what police would call a cold case review of the incident. This was essentially an analysis of the MoD file on the case, assessing what we knew and—more important—seeing what the investigators had missed.

The series of events began in the early hours of December 26, when duty personnel reported seeing lights so bright that they feared an aircraft had crashed. They sought and obtained permission to go off-base and investigate. They didn't find a crashed aircraft—they found a UFO.

The three-man patrol from the 81st Security Police Squadron—Jim Penniston, John Burroughs, and Ed Cabansag—saw a small metallic craft moving through the trees. At one point it appeared to land in a small clearing. They approached cautiously and Penniston got close enough to see strange markings on the side of the craft, which he likened to Egyptian hieroglyphs. He made some rapid sketches in his police notebook.

Later on, because of the complicated legal and jurisdictional position of United States Air Force bases in the UK, police from Suffolk constabulary were called out to the site where the object had apparently landed. They conducted a brief but inconclusive examination and then left. But three indentations were visible in the clearing, and when mapped they formed the shape of an equilateral triangle. A Geiger counter was used to check the site and the readings peaked markedly in the depressions where the object—possibly on legs of some sort—had briefly come to Earth.

News of the UFO encounter spread quickly around the bases and came to the attention of the deputy base commander, Lieutenant Colonel Charles Halt. He was skeptical, but had the witnesses write up official reports, including sketches of what they had seen. Two nights later Halt was at a social function when a young airman burst in and ran up to the colonel. "Sir," he stammered, "it's back." Halt looked confused. "What?" he retorted. "What's back?" "The UFO, sir—the UFO's back." Halt remained skeptical but gathered together a small team and went out into the forest to investigate. He subsequently stated that he went out with no expectation of seeing anything; in his own words, he said that his intention was to "debunk" the whole affair.

But he never did. He, too, encountered the UFO, becoming one of the highest ranking military officers ever to go on the record about a UFO sighting. As he and his men tracked the UFO, their radios began to malfunction and powerful mobile "light-alls," brought along to illuminate the forest, mysteriously began to cut out. One piece of equipment that didn't malfunction was the hand-held tape recorder that the colonel took with him to document his investigation. The tape recording still survives, and one can hear the rising tension in Halt's voice and in the voices of his men as the UFO approaches: "I see it, too . . . It's back again . . . It's coming this way . . . There's no doubt about it . . . This is weird . . . It looks like an eye winking at you . . . It almost burns your eyes . . . He's coming toward us now . . . Now we're observing what appears to be a beam coming down to the ground . . . one object still hovering over Woodbridge base . . . beaming down."

At one point the tension in their voices almost seems to become panic as the UFO makes a close approach and fires light beams down next to Halt and his men. Following these events, Charles Halt wrote an official report of the incident and sent it to the Ministry of Defence. Although somewhat innocuously entitled "Unexplained Lights," his report described the first night's UFO as being "metallic in appearance and triangular in shape . . . a pulsing red light on top and a bank of blue lights underneath . . . The animals on a nearby farm went into a frenzy." He went on to detail the radiation readings taken from the landing site and set out the description of his own sighting.

Halt's report was received by the same Ministry of Defence section where, a little over ten years later, I would spend three years researching and investigating UFO sightings. The report went to my predecessors, who began an investigation. But they were hampered by a critical mistake that was to have dire consequences. For whatever reason—and it may have been nothing more than a simple typographical error—Charles Halt's report gave incorrect dates for the incident. So when the MoD checked the radar tapes, they were looking at the wrong days.

Looking at radar evidence is a critical part of any UFO investigation. There have been plenty of spectacular UFO sightings over the years, many correlated with radar. The MoD's comprehensive UFO files detail several such cases, including ones where RAF pilots encountered UFOs and gave chase—unsuccessfully, I might add. In the absence of any radar data that might confirm the presence of the Rendlesham Forest UFOs, the investigation petered out. Yet, as I was to discover years later, the UFO *had* been tracked, after all.

I spoke to former RAF radar operator Nigel Kerr, who had been stationed at RAF Watton at Christmas 1980 and had received a call from somebody at RAF Bentwaters. The caller wanted to know if there was anything unusual on his radar screen. He looked, and for three or four sweeps, something did show up, directly over the base. But it faded away, and no official report was ever made. It was only years later that Kerr even heard of the Rendlesham Forest incident and realized he might have a missing piece of the puzzle.

At the time, however, in the apparent absence of radar data to verify the presence of the UFO, arguably the most critical piece of evidence was never followed up: The Defence Intelligence Staff had assessed the radiation readings taken at the landing site and judged them to be "significantly higher than the average background." In fact, they were about seven times what would have been expected for the area concerned.

In reassessing the case during my review, I was disappointed by what I found. I discovered a series of additional mistakes that had fatally flawed the first investigation: failure to cordon off the landing site, search it with metal detectors, or take soil samples; delay in reporting the incident to the MoD; failings in information-sharing between the MoD and the USAF. If the investigation had been handled differently, we might know a lot more today about the strange object that landed. While delay and poor information-sharing were arguably human errors, the root of the problem was confusion about jurisdiction and whether the British or the Americans should lead the investigation. My own view is that both had jurisdiction but that the UK authorities had primacy and should have led. In fairness, the difficulties were compounded by the unprecedented nature of the incident. There was simply no standard operating procedure to cover a situation like this. I rechecked the assessment of the radiation readings, this time with the Defence Radiological Protection Survey, and they confirmed the original analysis.

I have spoken with the key witnesses in this complex case on many occasions. I am convinced that they are being truthful, and while recollections vary in some instances, this is to be expected given the time that has passed and given the fact that events occurred over several nights, with different people being involved at different locations. Indeed, I would be suspicious if everyone told exactly the same story, because in my experience this would suggest improper collusion between the witnesses.

But the simple fact that this is a multiple-witness event, where those involved are military personnel and where there is physical evidence, makes this one of the most significant UFO sightings ever.

The late five-star admiral Lord Hill-Norton, the UK's former Chief of the Defence Staff (the equivalent of the Chairman of the Joint Chiefs of Staff in the United States), though retired at the time, often asked me to brief him on the UFO phenomenon and draft material for him on the subject—a fearsome task for a middle-ranking government official. He was particularly outspoken on the Rendlesham Forest case and felt strongly that the MoD's line on the incident (that the events were of "no defense significance") was entirely unacceptable and at odds with the facts. In a letter that he wrote to a UK defense minister, which he copied to me, the admiral summarized his views on the case as follows:

My position both privately and publicly expressed over the last dozen years or more, is that there are only two possibilities, either:

a. An intrusion into our Air Space and a landing by unidentified craft took place at Rendlesham, as described. Or:

b. The Deputy Commander of an operational, nuclear armed, U.S. Air Force Base in England, and a large number of his enlisted men, were either hallucinating or lying.

Either of these simply must be "of interest to the Ministry of Defence," which has been repeatedly denied, in precisely those terms.

Project Condign

On May 15, 2006, under the UK's Freedom of Information Act, which is broadly similar to the U.S. FOIA, the Ministry of Defence published a formerly secret report on UFOs. Much information about UFOs had already been released, both at the National Archives and on the Ministry of Defence's website, but the release of this latest study was different and totally unprecedented. The study was classified "Secret UK Eyes Only" and only eleven copies of the report were ever made. It ran to over 460 pages and was given the code name Project Condign. Work started in 1996 and the final report was not published until December 2000.

Interestingly, the timescale is broadly similar to the semiofficial COMETA Report from France, which was initiated in 1995 and released in 1999. There was no link between the two projects, and the high classification and extreme sensitivity of the UK study precluded liaison with any other country.

The report represented an attempt to undertake a proper, in-depth scientific study that was going to look at all the evidence the MoD had amassed over the decades and come to a definitive view about the UFO phenomenon. My opposite number in the Defence Intelligence Staff, who was my main DIS contact and my gateway to this secretive organization, had first discussed this with me in 1993. Like me, he seemed intrigued by certain UFO cases in our files and our discussions about UFO aerodynamics and propulsion systems were like something from a

Star Trek script. Nothing was said openly, but when conventional explanations for some of the most compelling UFO cases were eliminated, fingers were pointed suggestively upward. And whenever the question of who was operating these UFOs was mentioned, the marvelous phrase "these people" was used. More often than not, these were private meetings between the two of us, at which no notes were taken. However, on one occasion my boss accompanied me and sat in silence for most of what turned out to be a particularly surreal briefing. "What did he mean by the phrase 'these people'?" he asked, in an exasperated fashion, on the way back to our own office.

But how were we going to get a study commissioned when so many of our colleagues thought the MoD should drop its UFO investigations altogether, as the United States Air Force had done in 1969? One of our tactics to pull this off was a simple linguistic sleight of hand: We banned the acronym "UFO." One mention of a "UFO," and people's prejudices and belief systems kick in, be they skeptics or believers; the term was too emotive and had too much baggage. So we devised "unidentified aerial phenomena" (UAP) as a replacement, and tried to use this in all internal policy documents, retaining the phrase "UFO" only for our dealings with the public.

It worked. With the term "UFO" having been quietly dropped, we pushed to get a study approved. To my surprise and delight, given some of the more skeptical voices in the department, resources were eventually obtained. I assessed the formal proposal when it arrived and recommended to my bosses that the study be commissioned; against my expectation, my recommendation was accepted. However, the project was subsequently delayed, and in 1994 I was promoted and posted to a different section. Accordingly, I played no part in the study and am certainly not—as has been alleged on the Internet—its anonymous author.

So what did we get? After four years and 460 pages of analysis, had we solved the UFO mystery? Well, no, we hadn't. What we got was a comprehensive drawing together of some existing research, coupled with some exotic new theories. "That UAP exist is indisputable," the Executive Summary states, before going on to say that no evidence has been found to suggest that they are "hostile or under any type of control." But by its own admission, the report has not provided a definitive explanation of the phenomenon: "The study cannot offer the certainty of explanation of all UAP phenomena," it says, leaving the door open.

One of the most contentious aspects relates to what the report refers to as "plasma-related fields." Electrically charged atmospheric plasmas are credited with having given rise to some of the reports of vast triangular-shaped craft, while the interaction of such plasma fields with the temporal lobes in the brain is cited as another reason why people might feel they were having a strange experience. The problem with this is that there's no scientific consensus here, and as a good rule of thumb one shouldn't try to explain one unknown phenomenon by citing evidence of another. In other words, you can't explain one mystery with another one.

The report also deals with flight safety issues. There are numerous UFO sightings involving pilots, and the Civil Aviation Authority (CAA) has records of some terrifying near-misses between aircraft and UFOs. In one such case, on January 6, 1995, a UFO came dangerously close to hitting a Boeing 737 with sixty passengers on board on its approach to Manchester Airport. The CAA commended the pilots for reporting the UFO, yet the official report states that both the degree of risk to the aircraft and the cause were "unassessable." Numerous RAF pilots have seen UFOs, too. I have spoken to many such witnesses, not all of whom made an official UFO report. Project Condign has an intriguing recommendation when it comes to such aerial encounters: "No attempt should be made to out-maneuver a UAP during interception."

The Public Informed . . . the Public Denied

When I joined the MoD in 1985, it was a closed organization with limited public and media interface. But the UK's Freedom of Information Act (FOIA) came fully into force in 2005, and the department I left in 2006, after a twenty-one-year career, was virtually unrecognizable from the one I'd known when beginning there over two decades ago. The section where I worked was now so busy dealing with FOI requests that this had taken precedence over the research and investigation that was done in my day. Few UFO sightings were investigated in any meaningful sense of the word, and most sightings elicited little more than a standard letter. If the witness was a commercial pilot or a military officer, the incident was at least investigated, but not to the extent that had previously been the case.

By 2007, the workload involved in dealing with FOI requests about UFOs on a case-by-case basis was becoming intolerable and I know that staffs were getting increasingly frustrated. Accordingly, because of this administrative burden, the MoD decided to proactively release its entire archive of UFO files. The French government had done so in 2007, and MoD officials hoped that the move would assuage accusations that the British government was covering up the truth about UFOs. Indeed, both the MoD and the National Archives expected that this would be a good news story about open government and freedom of information. The MoD confirmed to me in December 2007 that the final decision had been taken and I duly broke the story in the media.

The 160 files, some of them containing hundreds of pages of documentation, comprise tens of thousands of pages in all. Each page has to be considered for redaction to ensure classified information and personal data aren't released. The first batch of eight files was released on May 14, 2008, and within a month there had been around two million downloads from the National Archives website. So far, many of the UFO sightings detailed are mundane, but there are some extraordinary accounts by civil and military pilots and sightings corroborated by radar evidence. The release program is expected to be completed in 2011.

The MoD was midway through its ongoing program declassifying its UFO files when it made the decision, in December 2009, to close its office for receiving reports from the public—the well-known UFO desk—much to the disappointment of many. I was surprised that there was no announcement in Parliament and no public consultation about the change in policy, which ended all correspondence with the British people about UFO sightings. Instead, the news was slipped out in a way designed not to attract attention, through an amendment to an existing document, "How to Report a UFO Sighting," in the Freedom of Information section of the MoD website. The new text stated that "in over fifty years, no UFO report has revealed any evidence of a potential threat to the United Kingdom" and went on to say that "MoD will no longer respond to reported UFO sightings or investigate them."

On the face of it, this looked like the termination of the MoD's UFO project, mirroring what had happened to Project Blue Book in the United States. But the real situation was subtly different. An MoD spokesperson told the press that "any legitimate threat to the UK's airspace will be spotted by our 24/7 radar checks and dealt with by RAF fighter aircraft."

This confirmed what I already knew: Behind closed doors, away from public scrutiny, the really interesting UFO sightings would not be ignored. Sightings from police officers, UFOs witnessed by civil or military pilots, uncorrelated targets tracked on radar—all these things will continue to be looked at, albeit outside of a formally constituted UFO project.

This should come as no surprise. After all, where evidence suggests that UK airspace has been penetrated by an unidentified object, this must automatically be of defense interest. Thinking and acting on a position of disinterest just because the intruding object is an unconventional aircraft would be dangerous. Like many countries, Britain remains vulnerable to espionage and terrorist attack. What if the "UFO" turns out to be a prototype spy plane or drone? What if it's a hijacked aircraft with its transponder turned off, as we saw on 9/11? Given the current security climate, this is *not* the time that we should be taking our eye off the ball, simply because of the baggage associated with the term "UFO."

I have mixed feelings about this recent and controversial development. On the one hand, cutting out the public seems a retrograde step in terms of accountability and open government, and perhaps even patronizing. On the other hand, UFO sightings in the UK were at a ten-year high, and the MoD was receiving more FOI requests on UFOs than on any other subject. Disengaging from this and concentrating on sightings from pilots and on uncorrelated radar targets may represent our best chance of making progress in our investigation of the UFO phenomenon. The reality is that UFOs will still be taken seriously and investigated. They will still be treated as something of potential defense significance, but unfortunately, now the general public won't necessarily be kept informed about these most important UFO cases.

While the MoD has been unnecessarily defensive concerning UFOs, constantly seeking to downplay the subject and the department's involvement, I have seen no evidence to suggest the existence of a conspiracy to cover up some sinister truth about UFOs. Most sightings are misidentifications of ordinary objects and natural phenomena. But there is compelling evidence in the MoD files and in the files of other countries to show that some UFOs can not be explained in conventional terms. While nobody has a definitive explanation for the UFO phenomenon, my research and investigations show not only that it exists but that it raises important air safety and national security issues.

Despite the extraordinary nature of some of the material in this chapter, everything I've written can be checked by referring to documents freely available at the National Archives or on the MoD website. However, people often ask me to go beyond the facts and into the realm of speculation. Never mind what I know, what do I think? What do I believe? How has my official work on the UFO phenomenon affected me? Twenty-one years of working for the government taught me to choose my words carefully.

In terms of my worldview, my government work on UFOs had a profound effect. Before I began my official research and investigations, I knew little about UFOs and had no particular beliefs about the phenomenon. Afterward, I felt that my eyes and my mind had been opened to a world that had previously passed me by. There was certainly more to the phenomenon than misidentifications or hoaxes. What of the 5 percent or so of sightings that defy conventional explanation? Could any of them be attributable to something exotic, or even extraterrestrial?

Many scientists now believe there must be life elsewhere in the universe. If there are civilizations within 100 light-years of Earth, the Square Kilometre Array, the world's most powerful radio telescope due to be completed in 2024, should be able to detect them. Could we have been visited by an extraterrestrial civilization? Several of my colleagues in the MoD, in the military, and in intelligence believed we have been, and I certainly can't rule out the possibility. If just *one* UFO turned out to be an extraterrestrial spacecraft, the implications are incalculable.

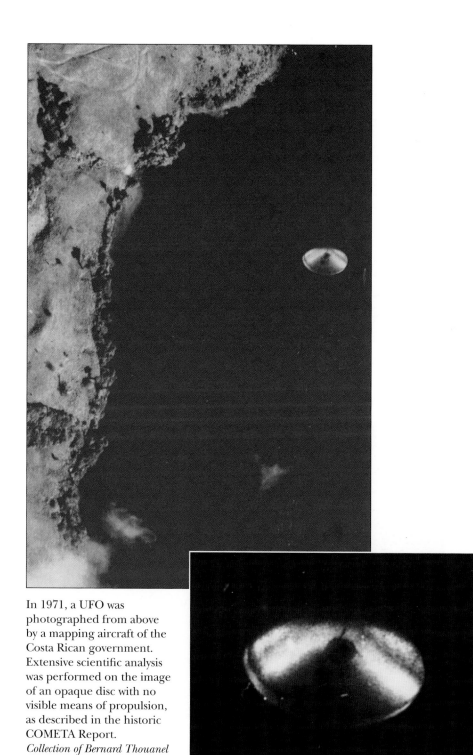

In 1971, a UFO was photographed from above by a mapping aircraft of the Costa Rican government. Extensive scientific analysis was performed on the image of an opaque disc with no visible means of propulsion, as described in the historic COMETA Report.
Collection of Bernard Thouanel

A rare photo of a UFO from the Belgian wave of 1989–90. The photographer was standing underneath the craft and captured four lights on its underside as it banked to the left. *© 2010, SOFAM/Belgium*

When the image was slightly overexposed, the object's triangular outline was clearly visible with a spotlight on each corner and a central light. *© 2010, SOFAM/Belgium*

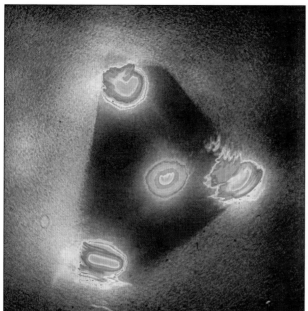

In one analysis, an expert highlighted a "halo" of light particles in a pattern around the craft, suggesting the presence of a strong magnetic field.
© 2010, SOFAM/Belgium

Taken by farmer Paul Trent with his wife in 1950, the classic pictures from McMinnville, Oregon, are among the most thoroughly analyzed in UFO history. The Trents were impeccable witnesses, and no evidence of a hoax has been found. *Courtesy of Dr. Bruce Maccabee*

In 1976, Iranian Air Force Major Parviz Jafari, now a retired general, was sent to investigate a brilliant, diamond-shaped object over Tehran in his F-4 Phantom jet.
Courtesy of Parviz Jafari

Peruvian Air Force pilot Oscar Santa María with his Sukhoi-22 fighter jet and a display of ammunition. In 1980, he was ordered aloft to attack a misidentified espionage device, which turned out to be a UFO.
Courtesy of Oscar Santa María

French Air Force Captain Jean-Pierre Fartek, a Mirage III pilot, had a close daytime sighting of a disc hovering near the ground.
Courtesy of Jean-Pierre Fartek

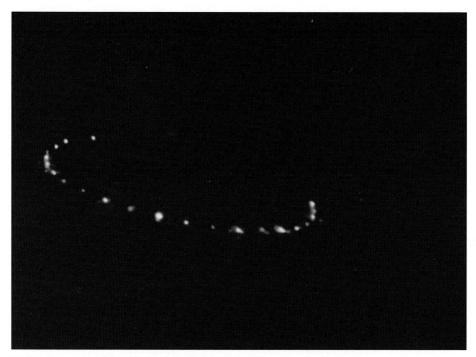

Lights from an unidentified craft photographed by a state police officer over Route I-84 near Waterbury, Connecticut, in 1987. Thousands of people witnessed UFOs in the "Hudson Valley wave" of the mid-1980s. *Collection of Phil Imbrogno*

Astronomer J. Allen Hynek, in the late 1970s. Hynek was scientific consultant to the U.S. Air Force's Project Blue Book for twenty years. At first a devout skeptic, he eventually recognized the physical reality of the UFO phenomenon.
Courtesy of J. Allen Hynek Center for UFO Studies

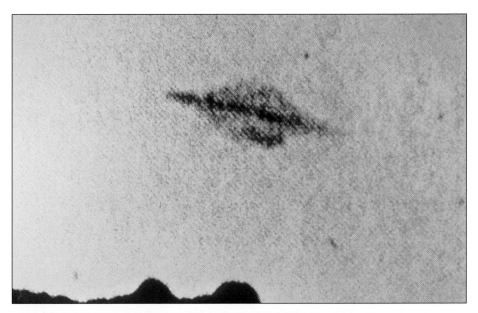

In 1958, a series of photographs was taken from a Brazilian Navy survey ship off Trindade Island, some six hundred miles east of Rio de Janeiro. Viewed by military officers and scientists on board, the object had a Saturn-like ring around it. The images were analyzed by the Brazilian Navy, and the president of Brazil vouched for their authenticity.
Courtesy of the Fund for UFO Research

Major General Denis Letty (Ret.) of France initiated a high-level private study of the UFO phenomenon, completed in 1999. I interviewed him at his home in 2008.
Copyright © Enzo Peccinotti

Left to right: Oscar Santa María, Anthony Choy, Ray Bowyer, Nick Pope, James Penniston, and Charles Halt at the National Press Club, 2007. *Copyright © Randall Nickerson*

John Callahan, former head of the FAA's Accidents and Investigations Division. *Copyright © Lisa Kimmell*

Fife Symington III, former governor of Arizona. *Copyright © Randall Nickerson*

Nick Pope, former head of the British Defence Ministry's unit investigating UFOs, at the Jodrell Bank radio telescope. *Copyright © Steel Spyda Ltd.*

Major General Wilfried De Brouwer (Ret.) was in charge of the Belgian Air Force investigation of that country's UFO wave beginning in 1989. *Courtesy of Wilfried De Brouwer*

Four-star Brigadier General José Carlos Pereira (Ret.), a commander of the Brazilian Airspace Defense Command from 1999 to 2001, retains access to his country's high-level UFO reports. *Courtesy of José Carlos Pereira*

CHAPTER 18

The Extraordinary Incident at
Rendlesham Forest

by Sergeant James Penniston (Ret.), U.S. Air Force, and by
Colonel Charles I. Halt (Ret.), U.S. Air Force

I. Sergeant James Penniston

In 1980, when I was twenty-five years old, I was assigned to the largest tactical fighter wing in the Air Force, the RAF Bentwaters/Woodbridge complex in England. I was the senior security officer in charge of Woodbridge base security. At the time, I held a top-secret U.S. and NATO security clearance and was responsible for the protection of war-making resources for that base.

Shortly after midnight on Christmas night—the early morning of December 26, 1980—Staff Sergeant Steffens briefed me that some lights had been seen in Rendlesham Forest, just outside the base. He informed me that whatever it was didn't crash . . . it landed. I discounted what he said and reported to the control center back at the base that we had a possible downed aircraft. I then ordered Airman First Class Edward Cabansag and Airman First Class John F. Burroughs to respond with me.

When we arrived near the suspected crash site it quickly became apparent that we were not dealing with a plane crash or anything else we'd ever responded to. There was a bright light emanating from an object on the forest floor. As we approached it on foot, a silhouetted triangular craft about 9 feet long by 6.5 feet high came into view. The craft was fully intact, sitting in a small clearing inside the woods.

As the three of us got closer to the craft, we started experiencing problems with our radios. I then asked Cabansag to relay radio transmissions back to Central Security Control (CSC), and he stayed back while Burroughs and I proceeded toward the craft. At first I was confused, not understanding what I was seeing. This was truly unbelievable. Then fear

179

struck me, but I told myself that I had to stay focused. Was this a threat to the base and to us? I had to determine that first and foremost.

When we came up on the triangular-shaped craft, there were blue and yellow lights swirling around the exterior as though part of the surface and the air around us were electrically charged. We could feel it on our clothes, skin, and hair. It felt like static electricity, which made your hair stand up and dance on your skin. But there was no sound at all coming from the craft. Nothing in my training prepared me for what we were witnessing. This was no type of aircraft that I'd ever seen before.

After ten minutes without any apparent aggression, I determined the craft was nonhostile to my team and decided to approach further. Following security protocol, we completed a thorough on-site investigation, including a full physical examination of the craft. After my first walk-around of the craft, astonishment and awe overwhelmed me. All fear was gone. During this process, I took photographs, made notebook entries, and relayed messages through Airman Cabansag to the CSC, following required procedures. The feelings I had during this encounter were like nothing I had ever known before.

On one side of the craft were symbols that measured about three inches high and two and a half feet across. These symbols were pictorial in design; the largest symbol was a triangle, which was centered in the middle of the others. They were etched into the surface of the craft. I

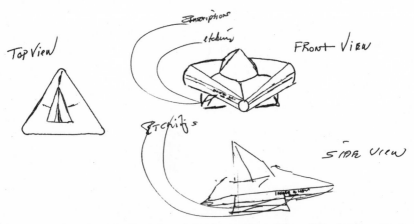

The drawing I made of the landed craft for the Air Force Office of Special Investigations. Collection of James Penniston

put my hand on the craft, and it was warm to the touch. The surface was smooth, like glass, but it had the quality of metal, and I felt a constant low voltage running through my hand and moving to my mid-forearm.

After roughly forty-five minutes, the light from the craft began to intensify. Burroughs and I then took a defensive position away from the craft as it lifted off the ground without any noise or air disturbance. It maneuvered through the trees and shot off at an unbelievable rate of speed. It was gone in the blink of an eye.

In my logbook, which I still have, I wrote "Speed <u>Impossible</u>." I subsequently learned that other personnel based at Bentwaters and Woodbridge, all trained observers, had witnessed the takeoff.

At that moment, I knew that this craft's technology was far, far above what we could ever engineer. When it took off, I felt alone, knowing now what John and I knew. Suddenly, there was no doubt. I realized that it was 100 percent certain that we are part of a larger community beyond the confines of our planet.

My drawing of the symbols, based on my logbook entry and my recollection of their placement. Their texture was rough, like sandpaper. Collection of James Penniston

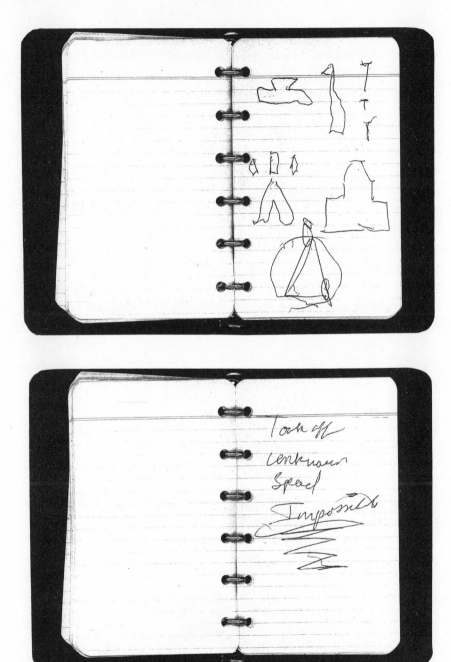

Two original logbook entries made while I observed the craft. Top: On one page I sketched the symbols. Bottom: When I watched the object take off, the speed was so shockingly fast that I wrote "Took off . . . unknown . . . Speed Impossible." Collection of James Penniston

After returning to CSC headquarters, we were debriefed and then advised to return to the landing site in daylight to look for physical evidence. After turning in our weapons and signing off, Burroughs and I went back and discovered broken branches scattered at the landing site. It appeared they had been forced down to the ground when the craft landed. There were scorch marks on the trees facing the site. But most importantly, we discovered three indentations in the ground, marks left by the UFO landing gear in the three corners of a triangle. I was relieved to find proof that this had really happened. I took photos of the landing site and, along with the ones I had taken of the UFO, took the film to the base laboratory. After taking Burroughs home, I went back to the site alone and made plaster casts of the three indentations left on the ground by the object.

The information acquired during the investigation was reported through military channels, and my team and other witnesses were told to treat the investigation as Top Secret. No further discussion was allowed. We were debriefed by First Lieutenant Fred Buran, on-duty shift commander at CSC; Master Sergeant J. D. Chandler, flight chief; and day shift commander Captain Mike Verano. In the days following, additional interviews were conducted by Colonel Charles Halt and then the Air Force Office of Special Investigations (AFOSI). This was a very hard time for me, as I was in shock by what I had witnessed.

I went back to the base photo lab, since I was the one who filled out the work order for the development of the two rolls of 35 mm film taken of the craft and landing site. I was told that the photos were apparently overexposed or fogged, and that none of them had come out. However, Senior Master Sergeant Ray Gulyas took six photographs of the site forty-eight hours after the event, which he developed off base and which survived; two of them show a British police officer and Captain Verano examining the site, and the three indentations are clearly marked with upright sticks.

I am still not sure about everything that may have happened during that night back in 1980. This event and its implications still weigh heavily on me. When all pieces of the puzzle are finally put together, then, hopefully, we can put the whole thing to rest. Until then, I will keep trying to find answers to the many questions that remain.

II. Colonel Charles I. Halt

In 1980, I was the deputy base commander of RAF Bentwaters, the large twin-base complex in East Anglia, England. As such, my duty was to provide support and backup for the base commander and to act as commander in his absence.

In late December 1980, I was called upon to investigate a strange event that was distracting our security police from their primary duties. Just after midnight, very early on December 26, 1980, our police patrolmen discovered strange lights in the forest east of the back gate of RAF Woodbridge. Three patrolmen—Staff Sergeant James Penniston, Airman First Class John Burroughs, and Airman First Class Edward Cabansag—were dispatched into the forest to investigate. They reported discovering a strange triangular craft sitting on three legs. The craft was about ten feet on each side, with multiple lights. It rapidly maneuvered and quickly left the area.

I was not immediately aware of the details, only being told of strange lights, and assumed there was a reasonable explanation.

Two nights later, the family Christmas party held on the 27th was interrupted by the on-duty police commander. He told of strange events and claimed that "it" was back. Since my boss had to present awards, I was tasked to go out and investigate. I fully expected to find an explanation.

I grabbed my pocket tape recorder and a cassette tape, and took four others with me into the forest: Bruce Englund (flight commander), Bobby Ball (flight superintendent), Monroe Nevilles (disaster preparedness NCO), and another young security policeman, Adrian Bustzina. John Burroughs, who had witnessed the event of two nights ago with Jim Penniston and was off duty, hitched a ride out and kept calling me on a borrowed radio. Neither he nor any of the other security policemen (at least fifteen or twenty) were allowed to come forward past the forest service road where the trucks and light-alls—motor-generated portable lighting systems—were parked. I was really upset that so many cops were out in the forest. It was a public relations nightmare just waiting to happen.

We went to the site where something had landed, and found the three indentations 1.5 inches deep and approximately 12 inches across on the ground in a triangular pattern. We took readings and discovered mild radiation and physical evidence, including a hole in the tree canopy above and broken branches. There were abrasions on the sides of trees facing the landing site. While documenting this examination by speaking into my tape recorder, I noticed some very strange sounds, which I thought were the nearby farmer's barnyard animals. "They're very, very active, making an awful lot of noise," I recorded on the tape.

Only seconds later, one of my men first observed a bright red-orange oval object with a black center in the forest. It reminded me of an eye and appeared as though blinking. It maneuvered horizontally through the trees with occasional vertical movement, zigzagging around the trunks as if under intelligent control. Here's an excerpt from my tape recorder as I watched, with some agitation:

Lt. Colonel Halt: We just bumped into the first light that we've seen. We're about 150 to 200 yards from the site. Everything else is just deathly calm. There's no doubt about it, there's some kind of strange flashing red light ahead.

Sgt. Nevilles: Yeah, it's yellow.

H: I saw a yellow tinge in it, too. Weird. It appears to be making a little bit this way?

Nevilles: Yes, sir.

H: It's brighter than it has been . . . It's coming this way. It's definitely coming this way.

Sgt. Ball: Pieces are shooting off!

H: Pieces of it are shooting off.

Sgt. Ball: At about eleven o'clock.

H: There's no doubt about it—this is weird!

When approached, it receded silently into the open field to the east. We watched in amazement for a minute or two. I recorded more on the tape:

H: Strange. One again left. Let's approach the edge of the woods at that point. Can we do without lights? Let's do it carefully, come on . . . Okay, we're looking at the thing, we're probably about two to three hundred yards away. It looks like an eye winking at you, it's still moving from side to side and when we put the star scope on it, it's sort of a hollow center right, a dark center, it's . . .

Lt. Englund: It's like a pupil . . .

H: It's like the pupil of an eye looking at you, winking . . . and the flash is so bright to the star scope, err . . . it almost burns your eye.

The reflection from the object flickered brightly on the west windows of a farmhouse across the pasture, on the side facing us, and I was concerned for the residents' safety. We could see the Orford Ness lighthouse farther to the right and a mile or so away, on the far side of the farm house, throughout the event.

Suddenly, the object exploded into five white lights that quickly disappeared. We went into the field and looked for residue, but found nothing. We then observed several objects with multiple red, green, and blue lights in the northern sky, which changed in shape from elliptical to round and moved rapidly at sharp angles. Several other objects were seen to the south and one approached at high speed, and then stopped overhead. It sent down a concentrated white beam—a small, dense pencil-like beam, like a laser beam—very near to where I was standing. It illuminated the ground about ten feet from us, and we just stood there wondering whether it was a signal, some type of communication, or maybe a warning. We really didn't know. The beam switched off, and the object receded, back up into the sky. I reported on this, once again, into my pocket tape recorder.

An object also sent down beams that night near or into the weapons storage area. I was several miles away, but we could see a few beams, and they were reported on the radio from the location. Later, others from the weapons storage area told me they had seen the beams. That caused me a great deal of concern. What was it doing there?

The whole time we had difficulty communicating with the base as all three radio frequencies—command, security, and law enforcement— kept breaking up. This activity continued for about an hour. During this entire event, I taped the various sightings as they unfolded on my pocket tape recorder, turning it on and off, and accumulated about eighteen minutes of recorded information.

The day after the incident, I ran into Colonel Gordon Williams, Wing Commander of the 81st Tactical Fighter Wing at RAF Bentwaters, in the common hallway. He had heard my radio transmissions the night before, and I played the tape for him. He asked to borrow it and took it to the Third Air Force staff meeting, where he played it for the staff, and for his boss, General Robert Bazley.

Williams told me that nobody had any ideas at the meeting, and they responded with silence. But he instructed me to contact the British RAF liaison officer, Don Moreland, stating that since this happened off base, General Bazley had declared it "a British affair." It turned out Don was on vacation, but when he returned, he asked me to file a memo (his absence explains the delay in the date of the document). I wrote up the details in my January 13, 1981, memo, "Unexplained Lights," and a copy was sent to the British Ministry of Defence and to the Third Air Force. The memo described the sighting of Penniston and the two patrolmen of the triangular object on the ground; the depressions and other physical evidence we found at the landing site; and the various lights and objects that I and numerous others witnessed subsequently.

Sometime later, my new boss found my tape and, unbeknownst to me, started playing it at cocktail parties. Word of it got out, and an American researcher started digging for more information. In 1983, I got a call from Pete Bent, acting Third Air Force Commander, and he told me that my memo from the Third Air Force files was going to be released under the Freedom of Information Act. I knew Pete and asked him to please burn it, to destroy it, and told him my life and his would never be the same because of what would happen if this were released. He said that too many people knew of the statement, so he had no choice. In October 1983, my worst fears were realized: The popular British tabloid *News of the World* ran a huge headline with the story on its front page, and reporters were swarming the base looking for the author of the memo. Fortunately I was already on a flight to the United States at the time, but this was only the beginning. In 1984, the audio tape was made public as well. The original tape was returned to me, and I also have the actual pocket tape recorder that was used that night.

If the memo had not been released, I would have continued to remain silent. This experience is not something I ever wanted to speak about publicly. On the other hand, no one has ever tried to influence me *not* to do so. When I had my final debriefing before leaving the Air Force, it wasn't even mentioned, so I asked if I could talk about the case, and was given permission, as if it really didn't matter.

Over the years, I have heard privately from many other witnesses. The weapons storage area tower operator and a communications worker in the same tower both told me they saw an object that went into the forest near Woodbridge base. The air traffic control tower operators at Bentwaters also saw an object and observed something cross their screen at extremely high speed, up to 3,000 to 4,000 mph—the radar monitor registered one streak as opposed to the usual series of blips for even the fastest aircraft. Others have now come forward with similar accounts. All had been cautioned not to talk by someone up the chain of command, or were afraid to talk at the time for various reasons.

Many have wondered how much the United States government might know about the Rendlesham Forest incident. Over the years, it has become clear to me that agents from the Office of Special Investigations (OSI), the Air Force's major investigative service, were on the base and secretly investigated the case in the days following. The incident had everyone very nervous. The high-ranking officers wanted to stay out of it, and the OSI didn't want anyone involved whom they couldn't control. OSI operatives harshly interrogated five young airmen, some of them in shock at the time, who were key witnesses. These men reported later that the agents told them not to talk about the UFO events, or their careers would be in jeopardy. Drugs such as sodium pentothal, often called a "truth serum" when used with some form of brainwashing or hypnosis, were administered during these interrogations, and the whole thing has had damaging, and lasting, effects on the men involved.

Other witnesses may have been exposed to high doses of radiation from the landed object. Some have health problems and struggle with personal issues to this day. Repression by the OSI is not uncommon in the military, but nobody involved will ever admit that. The OSI commander for Bentwaters told me then that they weren't investigating. Others have reported a different story.

I retired from the U.S. Air Force in 1991 as a colonel. This publicity was not exactly career-enhancing; nevertheless, I went on to become a base commander at two large installations and at the time of my retirement was director, inspections directorate for the DoD Office of Inspector General. In that position I had inspection oversight of all military services and defense agencies.

I still have no idea what we saw that night. It must have been something beyond our technology, judging from the speed of the objects, the way they moved and the angles they turned, and other things they did. I do know one thing, without a doubt: These objects were under intelligent control.

CHAPTER 19

Chile: Aeronautical Cases and the Official Response

by General Ricardo Bermúdez Sanhueza (Ret.),
Chilean Air Force, and by Captain Rodrigo Bravo Garrido,
Aviation Army of Chile

Along with West European nations, South America has also played a crucial role in establishing new agencies to investigate UFOs, and these endeavors have gathered some momentum in that part of the world. Peru set up its Air Force Office for the Investigation of Anomalous Activity, known as the OIFFA, in 2001, primarily concerned with the safety of air operations. And the Peruvian government took another important step about two years later. The Air Force publicly reported on its investigations of a series of sightings, videotaped in the remote Chulucanas area, stating that whatever was being seen was physically real but could not be explained. Announced in February 2003 by Peruvian Air Force colonel José Raffo Moloche, official acknowledgment of the existence of UFOs had never been provided publicly before by the Peruvian government, so this was an important breakthrough.

Comandante Julio Chamorro, the founder and first director of the OIFFA, had been previously stationed at La Joya air base and was a witness to the incident involving Oscar Santa María Huertas in 1980, when the base was put on alert. He told me that Peru funded its Air Force agency because "these anomalous events had occurred frequently enough over national territory to create a danger, and we recognized that they needed to be taken seriously." Chamorro says that, as director of the OIFFA, he had approached the U.S. Embassy on a number of occasions to discuss the situation for the purpose of requesting assistance, but received no response. "We have not been able to expect any help from the Americans in dealing with this problem," he says.

The Uruguayan Air Force, which has been active in UFO investigations for decades, declassified its UFO files in 2009 and made them public, including records of forty cases that remain unexplained, some involving military pilots. "The UFO phenomenon exists and I must stress that the Air Force does not dismiss an extraterrestrial hypothesis based on our scientific analysis," said Colonel Ariel Sánchez at the time, an officer with thirty-three years of active service who presides over an Air Force commission studying the cases.

Chile set up an agency within its civil aviation department, the equivalent of our FAA, in 1997 to investigate UFO cases affecting aviation safety. The CEFAA (the Committee for the Study of Anomalous Aerial Phenomena) was founded and directed by General Ricardo Bermúdez and soon developed a relationship with the aviation branch of the Chilean Army, through the work of Captain Rodrigo Bravo. Since leaving the CEFAA in 2002, General Bermúdez has prepared a graduate-level course in UAP for the University of Science and Technology in Santiago, "designed to furnish the students with the necessary tools for distinguishing between what is real and what is fictional regarding UFOs," as he describes it. He constructed the course to include a wide range of lectures by other professors in related fields, such as astronomy, space physics, and astronautics. In January 2010, General Bermúdez was reinstated as chief of the CEFAA, in an elaborate ceremony chaired by the director general of civil aviation. Representatives of the armed forces, police (carabineros), *and academic and research communities from all over Chile were in attendance, and the event was covered by the media. "It was a beautiful ceremony that had the full support of the authorities," Bermúdez wrote in an e-mail.*

I. General Ricardo Bermúdez Sanhueza

In the last days of March and beginning of April 1997, various anomalous aerial phenomena were observed over the city of Arica, in the far north of Chile. For two consecutive days, lights were seen west of the city and the airport, alarming the people of the region. Lights were also visible over the sea, apparently moving in a coordinated fashion. In addition to members of the civilian population, other witnesses included civil servants and official aeronautic experts at Aeropuerto Chacalluta, the airport in that city. The news made its way to the press, and the Ministerial Department of Civil Aeronautics, DGAC, issued a public statement acknowledging and confirming these observations. This was the first time the Chilean government had publicly recognized the existence of unidentified objects in national airspace.

Given the high profile of the case and the strong public interest in the subject, and discussions that had already occurred within the Air Force about addressing the UFO issue, General Gonzalo Miranda, the DGAC director, ordered the creation of a committee to study anomalous aerial phenomena. This group, the CEFAA, was charged with compiling, analyzing, and studying every incident involving anomalous aerial phenomena observed by any aeronautic personnel, civil or military. It began its work on October 3, 1997.

I was put in charge of the CEFAA from 1998 to 2002. As current director of the Technical School of Aeronautics, I had held other important educational posts in the Air Force, such as director of the School of Engineers and sub-director of the School of Aviation. I had been an active researcher of unidentified phenomena, especially when I served as aviation attaché to England. It was during that assignment that I came to the conclusion that there was something happening in the world's skies, and that we didn't know what it was. My position as director of the CEFAA demanded that I keep a scientific view on this topic, but it also meant that I was willing to consider any hypothesis about the origin and nature of these phenomena.

My duties were, among others, to head the regular sessions of the staff and members of the group, to guide the research efforts, and to provide the logistical framework for implementing those efforts. In addition, I promoted cooperation with university and scientific organizations, both national and foreign. These included working with Dr. Richard Haines and NARCAP, and the French government's GEIPAN. Every day I would check the progress of these various investigations and would oversee the design of their procedures. At times I carried on research myself, and was actively involved with case investigations.

Like America's FAA, the DGAC's legal mandate is to manage the national airspace and to ensure the safety of all civil, military, and commercial air operations. For the CEFAA, as well, working within this authority, aviation safety of commercial flights is the priority. Air operations demand careful preparation and execution, without any element of distraction for the pilots. The sighting of an unknown phenomenon is certainly a great distraction that could affect both the crew of the aircraft and the air traffic personnel in the control tower. It could potentially overload the radio communications for both the pilots and air traffic controllers if operators were to focus on the bizarre phenomenon, relaying details and questions, a fact that should concern the officials of any country. The policy of the CEFAA is to pursue solid cases with adequate scientific data, but only if there is an indication that the safety of the aircraft might have been at risk.

As the director, I stated early on that the CEFAA is committed to international cooperation for the following reasons:

- To share relevant information and new findings
- To provide an incentive for universities and scientific organizations to work on this in multidisciplinary teams in many branches of science
- To marginalize charlatans and pseudoinvestigators, and to denounce frauds
- To have a uniform method of investigative processes and analysis
- To coordinate recommendations for air traffic control operators when there are risks of electromagnetic effects or other hazards on board aircrafts

Chile has undoubtedly taken a great step forward in the investigation of anomalous aerial phenomena. And just as the Chilean Air Force was one of the first to be created in the world, it is also historic that we are also one of the first to officially recognize these phenomena and to form a government agency specifically for their investigation.

The CEFAA's official position has always been to recognize that something is happening in our skies, but we, as yet, do not know what it is. A large percentage of reports we've received have upon investigation been confirmed to be planets, meteorites, or weather phenomena, or do not provide sufficient data for analysis. Occasionally we are unable to make a ruling because witnesses refuse to be interviewed, are not credible, or are even committing fraud. Sometimes our pilots are afraid of ridicule, although that problem is improving somewhat. Of all the cases that have been analyzed, about 4 percent have no explanation, meaning that using all the technical means available, we cannot reach a satisfactory conclusion.

We believe there is a possibility that we will be forced to confront greater interference from UFOs in the future, especially considering the documentation of incidents by experts in other countries. We believe it is of utmost importance to be prepared.

Officially, Chile has not directly requested the cooperation of the U.S.A. However, in April 1998, the CEFAA informed the adjunct aeronautic official at the U.S. Embassy in Chile about our existence and mentioned Chile's interest in working with the appropriate agency in the United States to share experiences, policies, procedures, etc., regarding this topic. In July 2000, the CEFAA sent the embassy a document asking to consult the Pentagon about whether a sighting witnessed by a large number of people along the central coast of Chile the previous February had been due to activity by the National System of Antimissile Defense. Both requests went unanswered. To be frank, we've had no response from the United States any time we've tried to enlist its cooperation.

Now, as of early 2010, I have returned to my post as director of the CEFAA. We have three full-time investigators and many new cases to study.

In summary, I am convinced that UFOs exist and are a reality that cannot remain unacknowledged by governments. The phenomena are evident in all parts of the world and no efforts in their study should be neglected. Toward this end, international cooperation is vital to generate standards for protocols and policies for data analysis. Personally, according to my best judgment, I am in agreement with the findings of the French COMETA Report: There is a high probability UFOs are of extraterrestrial origin. However, until that hypothesis can be either confirmed or disproved, we should abstain from falling into the domains of either philosophy or religion. On the other hand, we should not discard that hypothesis just because it may sound harebrained. We need to put it through rigorous scientific analysis so we can come to viable conclusions.

Thirty-three-year-old Captain Bravo is the youngest of our contributors, and the only one currently on active duty in the military. I had a chance to spend a few days with him in late 2007, when he spoke at our Washington press conference with the permission of the Chilean authorities. Though Captain Bravo has never had a UFO sighting himself, he has become a meticulous investigator of pilot reports and an authority on this subject in his country.

II. Captain Rodrigo Bravo Garrido

Since the beginning of Chilean history there have been reports of unidentified phenomena, sometimes called UFOs, observed in our skies. Over the years, we have increased our capacity to explain many of the sightings, but there continue to be others without a scientific or logical explanation. In 1997, defense-related analysis was conducted within the telecommunications industry that touched on the issue of anomalous phenomena and their effects on electromagnetic fields. Cases were noted where there was a blockage of radio communications concurrent with the presence of a UFO near an aircraft.

Recognizing the potential impact on aviation safety, in October 1997 the Department of Civil Aeronautics, which is under the direct oversight of the chief commander of the Chilean Air Force, set up the Committee for the Study of Anomalous Aerial Phenomena, known as the CEFAA. In cooperation with aeronautic specialists, this agency handles the solid, well-documented reports of unidentified aerial phenomena.

In 2000, at age twenty-four, I was in training to become a military pilot. For my thesis, submitted the following year, I was assigned to pursue research into anomalous aerial phenomena in order to determine their effect and impact on aerospace security. The aviation branch of the Army of Chile, BAVE, had in its files many reports from military pilots describing incidents during flight involving aerial phenomena that did not correspond to normal air traffic. These incidents posed a potential threat to air safety.

One of our most important civil aviation cases occurred in 1988 and showed that unidentified flying objects can be a danger for air operations. A Boeing 737 pilot, on a final approach to the runway at the Tepual Airport in Puerto Montt City, south of Santiago, suddenly encountered a large white light surrounded by green and red. The light was moving toward the airplane, coming straight at it, and the pilot had to make a steep turn to the left in order to avoid a collision. The phenomenon was also observed by the control tower personnel.

More recently, in 2000, the crew of a Chilean plane from the aviation branch of the Army, flying south of Santiago, observed a long cigar-shaped object, a brilliant gray. It flew parallel to the right side of the aircraft for two minutes, very close to it, and then disappeared at an extremely high speed along the mountain coast. This object was detected by the radar of the control center of Santiago, which notified the crew minutes before the incident and confirmed the ensuing pilot observations.

It so happened that the pilot of this plane was the director of military aviation studies and was also my flight instructor during my training to be a military flier. Because of my connection to him, I had access to the full report on this incident, filed within my department by those involved, and I investigated the case further. I was able to interview the other pilots, the flight engineer, and the passengers on the flight who also observed the object.

In this unusual case, the military aviation crew members confirmed the reality of the UFO through careful observation and detailed reporting. Radar simultaneously confirmed the object's extraordinary movements; the case heightened official interest by Chilean military and aviation personnel in the UAP phenomenon. In fact, this significant event had a major effect on the attitudes and opinions of our military pilots. It was because of my involvement in this pivotal case that I was asked to study the unconventional topic of UAP in order to graduate from my pilot training program.

After conducting this investigation, I concluded in my thesis that UFOs are physically real and their presence in our skies is concrete. However, difficulties arise when we try to study their behavior, because of the complexity of the phenomenon and because of our inability to predict UFO events. I realized that the wide variety of different shapes, structures, colors, and movements of these UFOs meant that they must comprise a larger, more widespread phenomenon than we have understood.

When I became a pilot, I heard stories about encounters with unidentified flying objects, and became aware of the risks they could create, the possible dangers. In Chile there is excellent training provided in aerospace for all types of emergencies, but there is nothing written or taught about UFOs. This means that any reactions during UFO encounters are left up to the discretion of the pilots and will naturally have to be improvised on the spot. While I was engaged in UAP research, a link was established between our BAVE and the CEFAA, and the two agencies worked together to share information and cooperate on cases. Our objective analyses, and the serious regard with which this important phenomenon has been taken, have helped to foster UFO awareness among our flight crews. In cooperation with flight security they now maintain an openness about reporting any abnormal situation, and no longer ridicule the discussion of UFOs.

I have continued to study UFOs with the full support of the CEFAA for my studies of military cases and aviation issues, and cases from the Aviation Army are being sent directly to the CEFAA. To date, I have analyzed twenty-eight cases, nine involving aircraft of the Chilean Aviation Army. These nine well-documented cases have been studied by other officials in the Chilean government and were presented in official reports. We also collaborate closely with some civilian investigative organizations, which provide vast experience in research and which also exchange information with other countries.

Although my official position with the BAVE is not specifically concerned with UFOs, I am the individual whom pilots consult after an observation, before they submit their reports to their departments as required. I seem to be handling this issue more and more frequently, because it has become known that I am a point person for reports and investigations about UFOs.

At present, both the BAVE and the CEFAA are developing additional research methods and are creating an important database for future air operations. We do not keep this information secret. There is real interest in the subject of UFOs. But unfortunately, science does not support experiments or a testing of the evidence, and our current scientific methodology for measuring and verifying data is not easily applied to the study of the UFO phenomenon. As a result, the study of UFOs has attracted too many self-taught investigators promoting unscientific theories that are covered by the mass media. Because of this, in Chile as in other countries around the globe, UFOs are considered to be something standing apart from classical science and are rejected by established scientific institutions. So this makes it very hard to identify these anomalies existing throughout the world's skies.

Personally, I believe that the UFO phenomenon is the most interesting of all the many phenomena affecting our planet, and it is one that totally defies logical explanation. As of now it seems beyond our ability to comprehend. But new cases continue to be documented by pilots, air traffic controllers, operations staff at the world's airports, and many others with the proper training to determine whether a flying object is something unusual. Even though the true origin of these UFOs remains unknown, they do affect aviation everywhere, and this must be addressed. Eventually, I believe, we will be able to determine the real nature of this phenomenon if the scientific method is applied.

CHAPTER 20

UFOs in Brazil
by Brigadier General José Carlos Pereira (Ret.)

Most North Americans are not aware that Brazil is the fifth-largest country in the world, occupying most of the eastern continent of South America. It has spawned many dedicated UFO researchers and field investigators over the decades, earning it the dubious reputation of being a "hotbed" of weird UFO events. It also has a rich history of official involvement and Air Force reports. The Brazilian military has been investigating UFOs for many years, as shown by government documents.

For example, Brazil made a significant contribution in its release of one of the most important series of photographs in UFO history. Only a few clear UFO photographs have been taken by official sources, subjected to extensive laboratory analysis that verifies their authenticity, and then released to the public. Four images from Brazil, known as the Trindade photographs, are among the best, most valuable photos ever taken. The Brazilian government was involved with the release of these photos over fifty years ago.

At around noon on January 16, 1958, a retired Brazilian Air Force officer, Captain José Teobaldo Viegas, and Amflar Vieira Filho, chief of a group of submarine explorers, were the first—among many officers, sailors, and others—to sight an unusual object from the deck of a Brazilian Navy training ship. Almiro Barauna, a professional submarine photographer on board, managed to take a series of successful pictures over nearby Trindade Isle, despite the commotion on deck caused by the throng of excited observers. Captain Viegas later stated: "The first view was that of a disc shining with a phosphorescent glow, which, even in daylight, appeared to be brighter than the moon." About the size of the full moon, "it followed its path across the sky, changing to a tilted position; its real shape was clearly outlined against the sky: that of a flattened sphere encircled, at the equator, by a large ring or platform. "

The Brazilian Navy Ministry endorsed the Trindade photographs. A report from United Press International stated that "Navy Minister Adm. Antonio Alves Camara said, after meeting with President Juscelino Kubitschek in the summer Presidential Palace at Petropolis, that he also vouched personally for the authenticity of the pictures." Kubitschek ordered them released to the public, and the House of Representatives demanded an investigation by the Navy, which compiled a report. The original photos and negatives were analyzed by both the Navy Photo Reconnaissance Laboratory and the private Cruzeiro do Sul Aerophotogrammetric Service, both confirming their authenticity. Later, civilian experts in America conducted further analysis.

It was only recently, in 2008 and 2009, that the Brazilian government began its release of numerous previously secret UFO files and stated that it would gradually release them all in groups by decade, one decade at a time. As of this writing, documents, photos, and drawings from the 1950s through the 1980s have been made public—more than 4,000 pages—many of them concerning the Air Force's "Operation Saucer," involving extensive military investigations of UFOs in the Amazon region in 1977.

A. J. Gevaerd, coordinator of the Brazilian Committee of UFO Researchers, a prominent civilian group, and his colleagues have been instrumental in bringing about the release of these government files. Gevaerd was also the first to interview four-star Brigadier General José Carlos Pereira (Ret.), the highest-ranked Brazilian official ever to speak out about UFOs. Brigadier General Pereira has contributed an original piece on the handling of UFO events in Brazil at the highest levels, including his personal thoughts about the phenomenon, for this book. At his request, some of the material included in his piece was excerpted from a transcribed interview with Gevaerd, while some of it was written specifically for this piece. All was translated from Portuguese.

The general begins his essay with a description of a spectacular series of sightings involving military pilots and radar on May 19, 1986, which has come to be known as "official UFO night in Brazil." It was not until late in 2009—after Brigadier General Pereira completed his piece "UFOs in Brazil"—that any documentation was made

*public about this case. The newly released five-page "Occurrence Report"
about the 1986 incident was written by the acting commander of the
Brazilian air defense command to provide the Minister of Aeronautics
with "information provided by the air traffic control and air defense, as
well as the interceptor pilots involved in this event." The once-classified
report states that radar readings on both the air defense system and
intercepting jets were recorded simultaneously, while, also simultaneously, the pilots observed the objects through the cockpit window. Such
an "achievement" is quite rare: to capture a UFO on ground radar and
airborne radar while pilots observe it, all at the same time. This is what
the Belgian Air Force hoped to accomplish through its launch of F-16s a
few years later, as described by General De Brouwer.*

*The document lists numerous common characteristics of the phe-
nomena recorded that night, such as sudden accelerations and decel-
erations, an ability to hover, and supersonic speeds. The objects were
observed as white, green, and yellow lights, and sometimes without any
lights at all. The official conclusion reads as follows: "It is the opinion
of this Command that the phenomenon is solid and reflects intelligence
by its capacity to follow and sustain distance from the observers, and
also to fly in formation, not necessarily manned."*

*Brigadier José Carlos Pereira was a commander of the Brazilian
Airspace Defense Command from 1999 to 2001, and he then became
General Commander of Air Force Operations until 2005. In that position, he supervised thirteen generals and 27,000 subordinates. Prior to
these positions, he had been a commander of several air bases in Brazil
and commander of the Brazilian Air Force Academy.*

On the night of May 19, 1986, an array of UFOs were spotted over southeastern Brazil, and the entire defense system was put on alert. The Air Force scrambled its most experienced pilots in F-5 and F-103 jets to intercept these objects. Colonel Ozires Silva, president of a Brazilian oil company, and his pilot, Commander Alcir Pereira de Silva, were flying an executive Xingu jet near Poços de Caldas heading to São José dos Campos, when radars in different locations showed twenty-one UFOs in the sky from São Paulo to Rio de Janeiro. Silva and his pilot saw one of them and chased it for thirty minutes—a fast-moving, bright red-orange light that appeared to jump from point to point. They were not able to gain on it and eventually had to give up their pursuit.

This was a situation in which numerous expert witnesses saw something and radar detected the same thing. Radar equipment can be affected by many different factors, and can present a false echo, but a false target appears very briefly and is easy to recognize because it disappears quickly. It's a different story when we have a regular trajectory to follow. Also, when we have more than one radar spotting the same target, we know it's serious. This equipment operates in different frequencies, so we have the correlation of independent readings from different sources. These data have nothing to do with human eyes. When, along with the radar, a pilot's pair of eyes sees that same thing, and then another pilot's, and so on, the incident has real credibility and stands on a solid foundation.

A few days after these sightings, Brazil's Minister of Aeronautics, Brigadier Octavio Moreira Lima, called a press conference to explain what happened. He revealed that six jets had been scrambled from Santa Cruz AFB and Anapolis AFB, and some of the pilots had made visual contact, while all objects were registered on radar. The minister promised an official report within the next thirty days, but for some reason he changed his mind about releasing it. This was probably for some political reason, or maybe fear of panic because at that time the thinking was that the population might panic, if they knew. But in the meantime, the pilots and controllers were not prohibited from speaking about it.

The events of that night were really amazing, and some of our simple questions have simple answers: Did the pilots see the phenomena? Yes. Did the radars spot them? Yes. Did Ozires and other military pilots see them? Yes. Did pilots in commercial aircraft see them? Yes. Do the times of the sightings correlate? Yes. Do the trajectories of the objects correlate? Yes; all of this was technically analyzed. So, did it happen?

Yes, it did happen.

Everything was spotted by *both* aircraft radars and the radars on the ground. On-board radars operate in a microwave band, which is very narrow, while ground radars operate in a much broader band, so there's no risk of confusion or mistaken correlation.

During this event, the military was not fearful of any sort of invasion. Jets armed with missiles took off and reached the objects in less than two minutes. These jets are always armed, but with peacetime armaments, consisting of two small missiles. If those objects were from an enemy country, they'd have been crushed that night. These pilots were highly trained and their radar capacities were increased to the maximum, which normally isn't required. Radars never operate at full capacity, in order to save energy and to prevent wear-and-tear on the equipment. But after the jets took off, the capacity was increased to a broader range. Communications never failed, and the country was suffering no threat whatsoever. The jets landed safely and the pilots returned unharmed. Mission accomplished!

I don't think that UFOs have made any real threat to national security, but we have to recognize that the current lack of knowledge about the subject is enough to raise suspicions, as it would about anything as seemingly advanced. So we then come to the very biggest of questions: What were those objects? No one knows. They were not foreign jets attacking. They were *unidentified flying objects*. And where are these objects now? Who knows? Were they captured? Not that we know. So here is where the problem of material evidence comes in, and we don't have it.

When I was a commander, these unusual sightings occurred about once a month and usually were of very short duration. I remember there were about two to three incidents per year of military pilots being sent up to intercept something unknown that appeared on radar. Our civilian pilots are not afraid to speak up, and they always do, because they don't want to lose their jobs for *not* reporting unusual events. The first thing they do when they see something strange is to call the controllers, because they have a huge personal responsibility.

A civilian aircraft is always in contact with air traffic control, and all of these operations in Brazil are linked to the Air Force and are of a military nature. When a commercial pilot says, "There's something going on here," the control center will immediately report it to the military operations center in that area, in case it is something serious. They will take some action regarding that fact and report to the air defense operations center, which is the superior body and the only one to oversee the whole country. Then the pilot or the air traffic controller will fill out a report; they know where to get the form—from any Air Force base or any traffic controlling office throughout the country—and they deliver the completed papers to any Air Force base.

Next, there's always an investigation after the pilot registers what he saw. As requested on the reporting form, he must report the direction, altitude, and speed of the object. We also need other details, such as the position of the sun compared to the aircraft at that time. The brightness of the object is also important, as well as the kind of clouds in the sky at the time. All these data are precious. The controllers are then able to check if some other aircraft crossed the path of this pilot, which could explain the event. An investigation will follow, and if they discover that no other aircraft was there and the weather was not a factor, we have a special situation. And all these things are easy to check when everything is spelled out in the initial report. We go on eliminating all possibilities until we are sure that there is no conventional explanation for the data, and then the report is securely filed.

Pilot reports that turn out to have a conventional explanation are eventually deleted, and someone from the Air Defense will inform the pilot that they found out what happened. If no explanation is found, the case is transferred to another folder, called the "Book of Flight Occurrences." All of these unsolved cases are kept there in those books, and one hopes that researchers will eventually be allowed to see them. They include serious reports from pilots and air traffic controllers—everything we cannot explain, everything that is held as secret, goes to those books. It's important to emphasize that this "Book of Flight Occurrences" contains cases that couldn't be explained even after analysis by experts especially assigned to this task.

When I was a commander at COMDABRA, the Brazilian Airspace Defense Command, from 1999 to 2001, all cases involving UFOs spotted by military pilots and by radars would came to my attention. I directly participated in an investigation of a UFO incident only once, although I had access to secret files and both official and unofficial reports. After leaving the military, I still had access to nearly all the information I desired on this subject.

I haven't followed what happened at the Air Defense over the last four years, but I know that we continue to receive reports. Even so, I want to mention something important. I believe that up to 90 percent of all sightings are never reported. Brazil is a huge country, and these reports are filed only where there is an airport or an Air Force base, and only by people who know how the process works. Civilians don't even know that these forms exist and are available throughout the country. I don't know the actual percentage of sightings that result in reports, but I think it must be tiny. So the number of reports that come to the knowledge of the military is almost insignificant.

It is a big step for a country to officially acknowledge the existence of UFOs, as France has done. But releasing information has not caused people to panic, and I don't think it would if more files were to be opened. No one fears transparency; instead people fear the lack of it. I think that from the moment the government opens the subject for debate, all the fear people have toward this subject will disappear. And if there's one country that never panics, it's Brazil. Quite the opposite; maybe we would even create a new samba theme in celebration.

How do we handle the existence of UFOs? The evidence shows that unexplained phenomena are occurring, and this leads many of us to believe in the presence of alien spacecraft visiting planet Earth. However, drawing conclusions about what these things are is dangerous, since we do not have enough knowledge to do that. I believe science has much more work to do in order to identify and explain the phenomenon. We need astronomers, meteorologists, aviation experts, astrophysicists, and many other scientists, because such an investigation must be jointly addressed by many specialists. In fact, this effort must engage the whole nation. The synergistic effect of knowledge is undeniable.

I'm a man devoted to science, a man with a scientific mind. If you present the hypothesis that extraterrestrials may be here and may be doing things that we can't understand, your idea runs contrary to conventional scientific reasoning. As far as we know, our own solar system does not contain life on any planet except Earth.

I'm basing my ideas on the knowledge we have *today*, achieved by science as it currently understands the universe. This is the caveat to be considered. If we assume only current knowledge, I am forced to reject every possibility of anyone coming from outer space to Earth. And it gets more complex if we go further, because Alpha Centauri, the nearest star, does not seem to have a planetary system. We move then to the portion of the universe astronomers call the "inhabitable zone," which is many light-years from Earth.

However, I would never assert that no other civilization could have advanced a million years ahead of us somewhere else. I humbly insist, therefore, that our current knowledge must be inherently insufficient for comprehending everything. After learning about UFOs while in the military, I became clear—in fact, certain—about the high level of ignorance we have regarding the universe, given the current stage of human scientific development. The UFO phenomenon has demonstrated that we have a lot more to learn about physics and other scientific areas. We don't yet have the final word within science, and, eventually, we will be able to understand what is now unknown.

Look at what happened over the mere last hundred years, with discoveries ranging from penicillin to the airplane. We humans left the ground for the first time in an airplane nearly 100 years ago and within only one century were able to reach the moon. In astronomic terms a hundred years is nothing, not even dust. Obviously, an advanced people would not use rocket engines like our spacecraft sent into space. If in one century and with our limited capacity we could achieve this, think about it: Where will we be a hundred or a thousand years from now?

I don't have a problem with philosophy entering into this discussion in attempting to address the issues we haven't been able to solve: who we are, where we came from, and where we are going. Since Aristotle, human beings have been asking these same questions and we still don't know the answers. The scientific investigation of the UFO phenomenon in combination with other subjects within science and philosophy might be a way to move toward those answers.

No institution has the right to close the door on the discussion of any matters, be they scientific, political, social, or religious—and that includes the study of unidentified flying objects, which I consider to be within the realm of science. I believe that not only Brazil, but also all socially and technologically developed countries, should set up governmental agencies to address this matter. The United States should certainly lead the way, since that country is and will remain the planet's greatest technological power, with a great ability to aggregate knowledge from other countries. And if it should be accepted that something is coming here from space, I think the United Nations should be responsible rather than leaving that task in the hands of individual countries.

PART 3

A CALL TO ACTION

"The only way to discover the limits of the possible is to go beyond them into the impossible."

ARTHUR C. CLARKE

CHAPTER 21

Fighting Back: A New UFO Agency in America

D espite the astonishing yet rational deduction that the extraterrestrial hypothesis should be considered to explain some UFOs, as our experts have just pointed out, governments have an aversion to addressing that point or its implications. They are not motivated to pool resources and find out if this hypothesis can be proved, ignoring the popular interest in the subject and its potential for revolutionary discovery. In fact, the discomfiting quality of the extraterrestrial hypothesis—again, we're only talking about a *theory*, not a fact—likely explains why many governments want to keep a safe distance from the whole messy business. The difficulty of researching something as evasive and unpredictable as UFOs is also a problem—though not an insurmountable one. The agencies that *are* attempting to face the challenge have accomplished a great deal, as was demonstrated in the previous section, but ultimately they lack the resources to fully resolve the UFO mystery on their own. Even after many decades of focused research in France, exploration of defense implications in the UK, and field investigations in the Brazilian Amazon (to take three significant examples), we still don't know what the objects actually are. In their respective countries, some government agencies continue to collect case reports and look into sightings, adding more data to the heap but not solving anything, as the rest of the world looks away.

When asked, most military officers who have been personally involved with UFO incidents refrain from interpretation or speculation, yet privately many have a keen, persistent interest in getting to the bottom of the problem. They want to know what it was they themselves have seen, or what their trusted military colleagues have encountered, and this desire does not diminish over time. These witnesses and insiders recognize the extraterrestrial, or maybe interdimensional, possibility; once you have observed one of these bizarre manifestations at close range, your mind is newly opened, through no choice of your own. Even those who were prior debunkers, who would have scoffed at the mere notion of a UFO, are forced to recognize the once inconceivable. They often feel isolated, afraid of ridicule, unsupported by the world around them. But collectively, they may be able to make a difference.

Credible witnesses and government investigators have documented thousands of compelling case reports and first-person accounts. We now have accumulated enough data to establish the reality of some kind of consistent physical phenomenon *without a doubt.* Still, the American government lags behind, refusing to acknowledge any of this, leaving us American citizens stuck in a perpetual stalemate.

How can we overcome this? In terms of finding a workable model, we can look to France's UFO agency as the mother of them all, because, as we have seen, its office within CNES has been diligently working on the problem for over thirty years, from a research perspective rather than a military one. By seeking knowledge purely for its own sake, the French have been open to a wide range of explanations for UFOs, as scientists should be. The historic COMETA Report of 1999 broke a barrier when its generals, admirals, and engineers, along with a former head of CNES, brought the issue into the military realm and declared with great authority that even though it had not yet been proved, the extraterrestrial hypothesis was the most likely explanation of the phenomenon.

Will we ever be able to find out, to the satisfaction of scientists in the world community, what UFOs are and where they come from? Is this something we, as a planetary society, would be capable of deciding to do? If so, we would have to be proactive, rigorously seeking a resolution to this problem, making it a priority. Alternatively, would we prefer to sit back and wait for the seemingly all-powerful flying objects to reveal themselves more fully to us? Nearly all of the most concerned, most credible, and most serious of the government and military officials I have talked to agree on three basic points, when it comes to moving the issue forward:

- that further scientific investigation is mandated, partly because of the impact of UFOs on aircraft and aviation safety
- that this investigation must be an international, cooperative venture involving many governments and transcending politics

- that such a global effort cannot be effective without the participation of the United States, the world's greatest technological power

We are locked in by the stifling UFO taboo, which has served to protect us from the deeper, underlying issues and even threats—both perceived and unconscious—inherent in the most basic acknowledgment of a shocking and unexplained physical phenomenon. Now we need to rattle that cage. In this section, we will explore these crucial political questions with the help of a former high-level FAA official, a former state governor, and, more theoretically and philosophically, two leading political scientists. Yet, the final determination about our country's potential role in the future will have to be decided by all of us.

Logically, the first step in moving toward a solution is the establishment of an office or small agency within the U.S. government to handle appropriate UFO investigations, liaison with other countries, and demonstrate to the scientific community that this is indeed a subject worthy of study. In order to achieve these goals, we must consider where—under what branch of government—the United States should create this modest "UFO office" to get the process started. Using other countries as a model, there are many options. Often it is the Air Force that handles these investigations, as we have seen in Belgium and Brazil, even though neither government had established a special department within the Air Force for this purpose. However, in both cases, the generals involved have stated that a specific unit tasked full time with UFO investigations would have greatly aided the process, and they advocate for that necessity. Perhaps America needs to open a new Air Force office, being extremely careful to avoid repeating the many mistakes of Project Blue Book. General De Brouwer of Belgium recommends that the Air Force be the location for the American agency, because it is responsible for airspace security and has the means to intervene if required. The work of the office, he adds, must be objective, open-minded, and transparent, and private civilian groups could assist in this effort.

Four specific agencies described previously—the GEIPAN of France, the CEFAA of Chile, the OIFFA of Peru, and the Ministry of Defence office in the UK—were set up in four distinctly different bureaucratic departments within each of their respective countries. The French agency was founded within the equivalent of our NASA, while the Chilean authorities established theirs within the equivalent of our FAA, stressing aviation safety. The Peruvian office is an Air Force agency, and the British UFO office resided within their Ministry of Defence, like our DoD, with a mandate to protect the defense interests of the UK. This diversity of both locations and emphases has much to teach us, showing that within our own country we have a number of structural options.

Many of our contributors, such as Jean-Jacques Velasco of France, Dr. Richard Haines of the United States, General Bermúdez of Chile, and Brigadier General Pereira of Brazil, stress the importance of establishing some kind of centralized database—"a serious global organization that is objective, connected to agencies around the world, and committed to respond in a scientific and responsible way to the larger questions raised by the UFO issue," as Bermúdez describes it. "Without this, we are stuck." Some have therefore proposed that the United Nations might be a logical focal point for the further study of UFOs, since the phenomenon occurs worldwide, transcending national boundaries. Theoretically that makes sense, but its effectiveness would be highly unlikely, given the many preoccupations and bureaucratic headaches of today's world body in a time of increasing danger and hardship.

However, at an earlier time, in a relatively simpler world, an approach *was* made at the United Nations for just this purpose. Seven years after Project Blue Book was shut down, J. Allen Hynek and others attempted to establish an international investigative body within the halls of the UN.

In 1978, Sir Eric M. Gairy, then prime minister of Grenada, proposed to the United Nations General Assembly that the UN establish "an agency or a department of the United Nations for undertaking, coordinating, and disseminating the results of research into Unidentified Flying Objects and related phenomena." With his associates Dr. Jacques Vallée and Lieutenant Colonel Larry Coyne, a U.S. Army pilot whose helicopter almost collided with a UFO in 1973, Dr. Hynek requested—in a UN hearing—that the United Nations provide a framework in which the many scientists and specialists around the world working on the UFO phenomenon could share their studies. He pointed out that UFOs had been reported in 133 member states of the UN and that there existed over one thousand cases where "there appears physical evidence of the immediate presence of the UFO. In significant numbers, these reports had been made by highly responsible persons—astronauts, radar experts, military and commercial pilots, officials of governments, and scientists, including astronomers."

Despite these concerns, State Department teletypes show that the United States delegation at the UN was dismissive of Gairy's effort, calling it a "blitzkrieg sales pitch" and attempting to prevent his resolution from ever passing. A confidential message sent to the U.S. Secretary of State from the UN mission made an "action request" seeking "instructions on U.S. position to be taken in this matter as well as desired level of visibility. Last year Grenada requested our support and Misoff had to scramble hard behind the scenes to water down the resolution and, in effect, delay a vote for one year. Another consideration is whether to issue a disclaimer on statements made by U.S. nationals on the Grenadian delegation."

Later, U.S. members conducted "negotiating sessions" with delegates from other missions, "in an attempt to arrive at a mutually acceptable compromise solution to the problem." The plan was devised to refer the Grenada resolution to the Outer Space Committee without a mandate to engage in a study. This would alleviate "the need to vote on a resolution and gamble on the results." Despite U.S. efforts to block the vote, the General Assembly eventually adopted a draft resolution submitted by Grenada. It all fell apart in 1979, when Gairy was ousted during an internal communist takeover.

Hynek had also informed the UN committee about a study inaugurated by CNES, the French national space center, carried out by scientists from many disciplines. He remarked that the resulting case studies were "exemplary and far superior to the previous studies in other countries . . . the implications for science and the public at large of this French investigation are profound." The official French government agency GEPAN had just been formed within CNES under the direction of Yves Sillard, as part of a natural and logical response to a scientific, space-related problem that needed more research. At the same time, efforts were also under way in America to create a new UFO investigation within our own national space agency, NASA. But in America, it wasn't so simple—even when the request came to NASA from the very highest office in the land: the president of the United States. Unbeknownst to most Americans, even President Carter could not get the publicly funded agency to look at the UFO evidence and see if perhaps, just maybe, an investigative body within NASA was warranted.

Carter had had his own UFO sighting in 1969, before he became governor of Georgia. In 1973, while governor, he filled out a two-page reporting form by hand, in response to a request from a civilian UFO research group. According to his report, he was just about to give a speech at a meeting in Leary, Georgia, on an early October evening. He and ten members of the Leary Georgia Lions Club watched a bright, self-luminous object, at times as large as the moon. For over ten minutes, it changed colors and "came close, moved away, came close and then moved away," and at other times stood still; then it "disappeared."

A year and a half after Carter's election as president in 1977, his science advisor, Frank Press, wrote to NASA administrator Robert Frosch recommending that NASA set up a "a small panel of inquiry" to see if there were any "new significant findings" since the Condon report. "The focal point for the UFO question ought to be NASA," Press wrote, and Frosch's initial response was enthusiastic. "A panel of inquiry such as you suggest might possibly discover new significant findings," he replied in September. "It would certainly generate current interest and could lead to the designation of NASA as the focal point for UFO matters." He suggested that NASA name a "project officer" to review UFO reports from the last ten years and make a recommendation. The White House concurred without delay.

The U.S. Air Force, which had publicly declared UFOs not worthy of investigation, seemed to have deeply rooted hesitations about the Carter administration's request that NASA initiate a new inquiry. Colonel Charles E. Senn, chief of the Community Relations Division at the Air Force, stated in a letter addressed to NASA's Lieutenant General Duward L. Crow, "I sincerely hope that you are successful in preventing a reopening of UFO investigations." There is no record to indicate to what extent this or any other pressure from the Air Force influenced developments within NASA in response to Frank Press's request on behalf of Carter. Some NASA employees had reservations as well.

After a fairly lengthy series of letters, memos, and inquiries made through various levels of the hierarchical NASA bureaucracy, the agency turned down the president of the United States in December 1977—

without giving a project officer a chance to review the accumulated data. Frosch said that NASA needed "bona fide physical evidence from credible sources . . . tangible or physical evidence available for thorough laboratory analysis" in order to do so. Due to the absence of such evidence, he said, "we have not been able to devise a sound scientific procedure for investigating these phenomena." Therefore, he proposed that no steps be taken to "establish a research activity in this area or to convene a symposium on this subject."

Dr. Richard C. Henry, a prominent professor of astrophysics at Johns Hopkins University, was then deputy director of NASA's Astrophysics Division and involved in the decision-making process. In a 1988 published essay, Henry takes issue with Frosch's claim of "an absence of tangible or physical evidence." He says there was an abundance of relevant evidence at the time, a situation that he, as head of the Astrophysics Division, was certainly aware of.

Henry says Frosch's statement denying the existence of a sound scientific protocol was simply false. "The National Academy of Sciences endorsed the Condon study of UFOs, and specifically endorsed their procedures (protocol). It hardly does for us to say no sound protocol is possible!" he wrote in a memo to NASA space science administrator Noel Hinners. "The point is that *to be meaningful* the protocol must cover the possibility that the UFO phenomenon is due in part to intelligences far beyond our own." Ironically, it was this very Condon report which set the negative tone within mainstream science and no doubt influenced NASA's flimsy rejection of Carter's scientifically based presidential request.

Clearly, NASA appears to be an unlikely home for an American UFO agency. But what about the FAA? This agency seems to play a very different role in relationship to UFOs than the civil aviation departments of western European and South American countries, despite its mandate to protect our skies. We must remember that in 2006, the FAA informed the pilots and other aviation witnesses to the disc hovering over O'Hare Airport that it was actually weather, even though the weather was quite normal, it was daylight, and all weather data was recorded through standard procedures. When pressed, the FAA went a step further and attributed the sighting to a hole-punch cloud—a specific and quite rare weather phenomenon that requires freezing temperatures to occur—despite the fact that the temperatures at O'Hare that afternoon were well above freezing. Such irresponsible statements serve to discourage witnesses from filing reports, which would normally be the first step in conducting any sort of investigation. Unfortunately, the FAA seems like an even less likely candidate than NASA to take on UFOs at this point.

A comparison with the Civil Aviation Authority (CAA) of our closest ally, the United Kingdom, is in order. There, it is *mandatory* to report any incident where pilots or aircrews believe there has been any danger to their aircraft—whatever the source. Then the CAA and other authorities have the basis upon which to decide if an investigation is warranted.

After Captain Ray Bowyer and his passengers observed a pair of brilliant objects over the English Channel in 2007, the first thing Bowyer did upon landing was fax a report to the CAA, following standard required procedure. There was no attempt by his airline or anyone else to hush up the story, which was reported by the BBC. In fact, many CAA files on unsolved cases involving pilots, air traffic controllers, and ground crews have been released. For example, in 1999, a BBC news item reported that "a UFO that narrowly avoided colliding with a passenger jet flying from London's Heathrow Airport has baffled aviation experts." A metallic object passed within twenty feet of the aircraft, but for some reason was not picked up on radar. The BBC reports that the pilot filed a near-miss report (an "airprox") and that "a report by the Civil Aviation Authority found no explanation for the incident which has also confounded local military experts and local police."

Imagine if the FAA had made such a statement about the equally radarless O'Hare incident. Being used to a saner approach, Captain Bowyer found the U.S. "non-reporting system" hard to imagine, because the CAA makes no distinction among the possible causes of distress on aircraft. How odd, upon reflection, that America's FAA seems to discount one rare hazard—unidentified flying objects—and recognizes all others, even if the potential impact could be the same. The FAA provides no reporting forms for these kinds of sightings—although it does offer report forms for volcanic activity and bird strikes, and a detailed "laser beam exposure questionnaire."

The FAA does not try to hide its discrimination. As a matter of policy, the agency has informed its employees that it wants nothing to do with reports of UFOs or anything anomalous, no matter how severe the danger to the aircraft or the lives within it. The 2010 FAA Aeronautical Information Manual, in Section 6 on "Safety, Accident, and Hazard Reports," states that "persons wanting to report UFO/unexplained phenomena activity" should contact a collection center such as Bigelow Aerospace Advanced Space Studies, a new research organization focusing on novel and emerging spacecraft technologies, or the National UFO Reporting Center (NUFORC), a civilian group with a UFO hotline and reporting forms that keeps careful records of UFO sightings.

With unintentional humor the manual goes on to say that "if concern is expressed that life or property might be endangered" by the UFO, "report the activity to the local law enforcement department." Does this mean the local police department over whose jurisdiction the jet is flying at the time it is endangered at, say, 35,000 feet above the ground? Or the nearest police force to an airport that might have a UFO hovering over it? Assumedly, such illogical directives would be changed if our country ever set up a UFO agency.

Two witnesses to the O'Hare incident did just as the manual suggested: They called NUFORC and submitted written reports on their sightings. Ironically, both told me they had never read the FAA manual and were unaware that the official tome dictated that this is what they should do! Both had heard of NUFORC independently and didn't know where else to go with their information, which they felt, as a matter of duty, needed to be on the record. It was these reports that were eventually provided to the *Chicago Tribune,* prompting transportation reporter Jon Hilkevitch to investigate further and eventually to break the O'Hare story on the front page.

It is my understanding that most FAA employees probably do not read the manual—certainly not cover-to-cover—but when sightings occur they seem aware of their employers' attitudes regardless. The message is conveyed to them, often subtly and indirectly as a kind of veiled professional threat, that they are not to talk to the press about these incidents. The FAA's negligence may border on the dangerous, or the problem may be that other government agencies need to take more responsibility for UFO incidents that the FAA claims are outside its jurisdiction. No matter which branch of government does so, the threat, if there is one, posed by unidentifiable objects in proximity to commercial aircraft needs to be properly assessed by a new unit established to investigate UFOs.

Nick Pope, former MoD official and UFO expert in the UK, says that governments define "threat" in a very specific way, especially within military intelligence circles. The formula goes like this: Threat = capability + intent. For example: the United States is aware that the UK has nuclear weapons (threat) and therefore could launch a nuclear attack on America (capability), but since the UK has *no intent* to launch such an attack, the United States faces no threat in this regard. Pope points out that we certainly know that UFOs have the capability of being a threat, given their fantastic speed and maneuverability, far superior to our own technology. But, in this case, the intent of UFOs is completely unknown, and therefore immeasurable. Because of that fact, UFOs must be taken seriously as possible threats, and the UK's Ministry of Defence monitors them for that reason.

Pope suspects that United States military intelligence circles also define "threat" in this way. The fact that the FAA instructs its employees not to report this particular potential threat lies in contradiction to this basic formula. Maybe it's time to change the FAA manual and provide employees with the proper reporting forms.

U.S. government reticence about addressing the problem of UFOs seems to have infected all departments that could potentially house a new agency for investigations. Yet we *can* overcome these obstacles through a rational, commonsense approach. Some authorities have suggested specific ways forward, based on their direct experience.

In the late 1980s, John J. Callahan was head of the FAA's Accidents, Evaluations, and Investigations Division, an extremely high-level position just one rank below federal positions appointed by Congress. When working with military agencies, Callahan's rank (GM15) was equal to that of general.

One day in early 1987, he was unexpectedly faced with the problem of managing a UFO case—a dramatic, thirty-minute sighting by three Japan Air Lines pilots of a giant UFO over Alaska. Previously, Callahan had never given the slightest thought to the subject of UFOs. When he first heard about the JAL case, he requested the extensive data be sent to him immediately and he brought it to the attention of FAA administrator Admiral Donald D. Engen. Admiral Engen set up a briefing, which, according to Callahan, included members of President Reagan's scientific staff, as they were described to him at the time. It also included three CIA agents.

Callahan did not say anything publicly about his role in the incident until 2001, thirteen years after his retirement. While talking to some close associates in his community who had probed him for information, he decided that it was time to speak out. The data from this case had been shipped to his home office when he retired, and had languished in his barn for all those years. A few charts were even nibbled on by mice, he discovered later. Fiery and blunt with a somewhat folksy style and a biting sense of humor, John Callahan makes no bones about the fact that he is not happy with the way the FAA conducts itself regarding UFOs. Nor is he in favor of withholding information about the subject from the public, and he's armed with the evidence, the experience, and the authority to make a very strong case.

So far, no one else has come forward who attended the debriefing at the FAA's Washington headquarters described by Callahan. I made a FOIA request to the FAA for Admiral Engen's log of appointments and schedule during this time, but was told no such records exist (Engen has since died). I called Callahan's boss at the time, Harvey Safeer, now retired in Florida. Safeer remembered the Alaska incident, but had no recollection of any such meeting taking place.

John Callahan's wife, J. Dori Callahan, was a major player at the FAA in her own right at the time of the incident. Initially an air traffic controller, Ms. Callahan was branch manager for Flight Service Data Systems (FSDS) of the Airways Facilities organization, the part of the FAA which provides the hardware support for all its air traffic control systems. She later became division manager for the Automated Radar Terminal Systems (ARTS) software programs, and retired from the FAA in 1995, after twenty-eight years there.

Dori Callahan remembers well that this high-level debriefing was called a short time after her husband presented his data to the admiral, and also that he told her what happened there immediately afterward. In addition, as an FAA expert, she later analyzed the radar printouts on the Alaska case, which Callahan had provided for the CIA at the meeting, along with the explanatory drawings prepared by the engineering and software staffs from the Tech Center. "And since I had worked in both hardware and software organizations at one time, I understood all of it," she explained in a 2009 e-mail.

John Callahan points out that, when looking at the unusual radar data during the briefing, the hardware department said it was obviously a software problem, and the software department said it was clearly a hardware problem. "Both teams were fully experienced and knew the air traffic software system, and both were fully capable of knowing when the system was not working correctly," Ms. Callahan stated in her e-mail. "In other words, there was nothing wrong with the hardware at the time of the JAL 1628 sighting, and the software was working as well. Looking at the radar display of the object darting in and around JAL 1628, it was obvious that there was an object changing positions around the jet. If it had been ghosting [a false target] as suggested by the FAA, all traffic in that control area would have had ghosting, and it would not have moved in front of and behind the aircraft."

In contrast to the O'Hare incident, the FAA *did* conduct an official investigation two months after the Alaska event—mainly because there *was* radar evidence, and because "public interest" forced the issue. The FAA wanted "to ensure that somebody didn't violate airspace we control," a spokesman explained at the time.

But maybe there were other reasons the agency looked into this. Despite the FAA's proclaimed disinterest in UFOs, Richard O. Gordon, an official from the FAA's Flight Standards Office, informed the JAL captain of a surprising scenario during a lengthy 1987 interview. He said that the captain's detailed account was "very, very interesting and we need to see if we can figure out what is there." As revealed in a verbatim transcript, Gordon then described plans to take the information provided by the captain and send it to Washington so authorities there could find out if it matched any previous reports. "We have a lot of stuff where pilots have had other sightings," he declared. He told the captain that maybe his description and drawings will be the same as what happened "in Arizona and New York or wherever," and that "we got a place in Washington, D.C., we'll put them all together" to find out if any two cases are alike. This is a very interesting admission: the FAA keeps records of UFO sightings by pilots; they're stored in a specific location in Washington, D.C.; FAA officials make case comparisons when new incidents occur. If it's true, it certainly flies in the face of the agency's public stance on UFOs.

Despite the reaction of individual FAA officials directly involved with the Alaska case, the stated FAA conclusion was that the radar readings were false targets, malfunctions in the system. Even though it had radar to support the witness accounts, the FAA dismissed this data as erroneous, and declared that it "was unable to confirm the event." It praised the three "normal, rational, professional pilots," yet the final report completely ignored the visual sightings reported in detail during the FAA's interviews with these witnesses.

John Callahan vigorously disputes these claims about the radar. He makes the important point that radar is not configured to detect objects that behave the way UFOs do, and that we need to revamp and upgrade its technology. This former head of the Accidents and Investigations Division was not at all surprised by the FAA's response to the O'Hare incident a few years ago. "It was predictable," he told me. "When pilots report seeing such an object, the FAA will offer a host of other explanations. It's like wearing a blindfold. It's always something else so it can't be what it is."

CHAPTER 22

The FAA Investigates a UFO Event "That Never Happened"
by John J. Callahan

You are about to read about an event that never happened.

I was division chief of the Accidents, Evaluations, and Investigations Division of the FAA in Washington from 1981 to 1988. During this time, I was involved in an investigation of an extraordinary event but was asked not to talk about it. Since retiring, I decided that the public had a right to this information, and that they could handle it. Nothing dire has occurred as a result of my discussing this incident publicly, yet nothing useful has resulted from it either, although it's never too late. I have come to realize the serious need we have to improve our radar systems so they can capture unusual objects in the sky, such as the one I dealt with when I was at the FAA in 1987.

It was early January 1987 when I received a call from the air traffic quality control branch in the FAA's Alaskan regional office, requesting guidance on what to tell the media personnel who were overflowing the office. The media wanted information about the UFO that chased a Japanese 747 across the Alaskan sky for some thirty minutes on November 7, 1986. Somehow, the word had got out.

"What UFO? When did this take place? Why wasn't Washington headquarters informed?" I asked.

"Hey," the controller replied, "who believes in UFOs? I just need to know what to tell the media to get them out of here."

The answer to that question was easy: "Tell them it's under investigation. Then, collect all the data—the voice tapes and computer data discs from both the air traffic facility and the military facility responsible for protecting the West Coast area. Send the data overnight to the FAA Tech Center in Atlantic City, New Jersey." I wanted the data on the

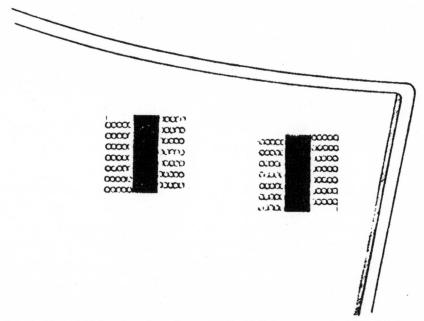

Captain Terauchi's drawing of two "spaceships" with light arrays or horizontal "exhaust" flames around a central object as seen through the cockpit window, provided to the FAA. Courtesy of Dr. Bruce Maccabee

midnight redeye flight, no matter how much hassle it was for them to get it to me.

Japan Air Lines flight 1628, a cargo jet with a pilot, copilot, and flight engineer, was north of Anchorage, and it was just after 5:00 p.m. The captain, Kenju Terauchi, described seeing a gigantic round object with colored lights flashing and running around it, which was much bigger than his 747, as big as an aircraft carrier. His crew, Takanori Tamefuji and Yoshio Tsukuda, both saw it, too.

At one point, two objects appeared to stop directly in front of the 747, and the captain said they were "shooting off lights," illuminating the cockpit and emitting heat he could feel on his face.

The objects then flew in level flight with the 747. Later, the captain made a turn to evade the UFO, but it flew alongside the jet, keeping a constant distance. Terauchi was able to estimate the size of the largest "spaceship," as he called it, to be at least the size of an aircraft carrier because he had it on his radar, and the aircraft radar has range marks. He reported all of this to FAA officials, exactly as he saw it.

Over the course of thirty-one minutes, the UFO jumped miles in merely a few seconds. One radar sweep at the air traffic control in Anchorage took ten seconds. At one moment Terauchi says, It's over here at twelve o'clock at eight miles, and when the radar antenna goes by, we see a target there. Ten seconds later, it's suddenly six or seven miles behind him. It's going from eight miles out in front of the 747 to six or seven miles in back, in only a few seconds, in one sweep of the radarscope. The technology was "unthinkable," Terauchi said, because the UFOs appeared to have control over both inertia and gravity.

FAA officials interviewed the captain and his crew extensively in the days and months following; all of them provided independent descriptions and drawings of the "spaceships" and their remarkable behavior. These three reliable witnesses knew how to recognize aircraft. If this object had been a secret military exercise, the pilots would have been informed as such and would not have wasted time spending thirty-one minutes evading and reporting a UFO, and the FAA would not have bothered to conduct interviews following the event. These witnesses eliminated *all* known explanations for what they had observed at close range for an extended period of time.

When a pilot looks out the window and sees an aircraft shooting across his nose or flying along with him, the first thing he does is call air traffic control and say, "Hey, do you have traffic at my altitude?" And the controller panics, looks at the scope, and says, "No, we don't have any traffic at your altitude." Air traffic would then question the 747 pilot asking for more information: what type of aircraft, any visible markings, color, or numbers on the tail, etc., and then the controller would advise, "We will track that guy and have flight standards meet him at the airport when he lands. We'll write him up; pull his ticket. We'll do whatever we have to do to find the pilot of the unknown aircraft." If his ticket was pulled, the pilot was no longer authorized to fly.

In this case, the pilot responded by saying, "It's a UFO," because he could see it so clearly. But who believes in UFOs? This is the type of attitude the air traffic control had at the time, and in any case, neither the controller nor the FAA was equipped to track something like this. The FAA has procedures that cover tracking unidentified aircraft, but it has no procedures for controlling UFOs.

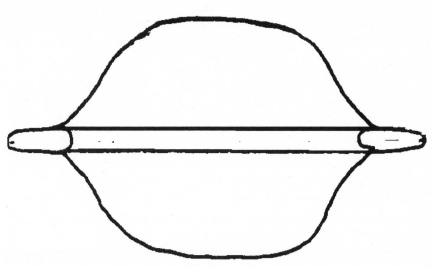

Captain Terauchi sketched a silhouette of a giant ship, which he said was the size of an air-craft carrier, with pale white lights on the horizontal rim. Courtesy of Dr. Bruce Maccabee

After receiving the call concerning the UFO from the Alaskan region almost two months after the UFO event occurred, I briefed my boss Harvey Safer, who alerted the FAA administrator Admiral Engen. Safer and I drove up to the FAA Tech Center in Atlantic City, New Jersey, to observe the computer playback of the event and learn more about what had happened.

The FAA had developed a computer program capable of re-creating the traffic on the controllers' scope, called plan view display (PVD). I instructed the FAA specialist to synchronize the voice tapes with the radar data—that way, we could hear everything the controller and pilot said, while simultaneously watching the radar scope. It would be just as if you were standing behind the controller in Alaska, watching everything that was going on while he conversed with the JAL pilot and crew. I videotaped the radar display as the event was played back.

Later that day, I asked the FAA automation specialists to plot the radar targets along the route of flight on a chart and explain what each target was doing along the 747's flight path.

The hardware and software engineers put together a large chart that showed every target along the flight of the 747 during its reported encounter with the UFO. They hung it on the wall and pointed out: This is when we first saw the UFO; this is when the pilot saw the UFO; this is when the military saw the UFO; all the way down the whole chart. I videotaped the chart.

The printout and radar playback displayed primary targets in the vicinity of the 747. These target returns were displayed about the same time and place as the pilot reported viewing the UFO. The pilot and crew viewed the target on their own radar and were able to actually see the huge UFO simultaneously, as it approached their aircraft. Anyone who watches this play back can see and hear this, but of course when the CIA saw it, their people said you can't see it because it's not there. The question I always ask is: Who are you going to believe, your lying eyes or the government?

Both the radar and manual controller observed the primary target. The military controllers also viewed the primary target on their radar and identified it as a "double primary," which means it was large enough to be more than one aircraft.

During the briefing at the FAA Tech Center in Atlantic City, I asked both the hardware and software engineers (these were the same people who had built the air traffic control system) to tell me what those dots in the vicinity of the JAL aircraft were. The hardware engineers said, "This target over here is a software problem, and this one over here is also a software problem." Every time, all the way down: It's a software problem; there is nothing wrong with our hardware system. So, I said, "Fine, that makes sense to me."

Then the software guy got up and said, "This target over here, it's a hardware problem and this one here—a hardware problem." There were no software problems, and there were no hardware problems. "Well," I asked, "what do we have if we don't have anything? Do we have a target there or not?" One of the technicians stated, "My religion forbids me to believe in UFOs," so I said, "Fine," and got ready to leave.

When I arrived back at FAA headquarters, I gave Administrator Engen a quick briefing of the playback and showed him the video of the radar scope synchronized with the voice tapes. He watched the full half hour, and then set up a briefing with President Reagan's scientific staff, and told me my function was to give them a dog-and-pony show and hand this operation off to them, "since the FAA does not control UFOs."

At the briefing, we looked at the data printouts and played the video for the people there two or three times—the participants turned out to be the CIA, the president's scientific group, and a bunch of grunts. We talked for an hour and half or so, and the scientists asked a number of questions—very intelligent questions, in fact. They wanted to know things like the speed of the radar antenna, the frequency and the bandwidth, and the algorithm for the height-finding equipment. The FAA people we brought into the room were technical engineers—hardware and software specialists—and they gave those answers like they were high school math coaches. They spit that stuff right out; it was really amazing to watch these FAA experts work.

At the end, one of the three people from the CIA said, "This event never happened; we were never here. We're confiscating all this data, and you are all sworn to secrecy."

"What do you think it was?" I asked the CIA person.

"A UFO, and this is the first time they have over thirty minutes of radar data to go over," he responded. They—the president's scientific team—were very excited to get their hands on this data.

"Well, let's get a Twix out and advise the American public that we were visited by a UFO," I suggested.

"No way. If we were to tell the American public there are UFOs, they would panic," he informed me.

And that was it. They took everything that was in the room—and in those days, computer printouts filled boxes and boxes. These FAA printouts were titled "UFO Incident at Anchorage, 11/18/86," written on the front cover. The printouts provided ample data for an automation specialist to be able to reproduce everything the controller saw on a chart.

A few weeks later, an FAA technician brought in the FAA's report of this event that never happened. I had him put it on a little table in the corner of my office, and said, "Leave it there. When the CIA wants the rest of the data, I'm sure they'll come and get it." Some time passed and someone brought in the voice tapes from the incident, and we put that next to the report on that table, waiting for the CIA to come and make a pickup.

The chart produced at the Tech Center also came to my office, where it remained for a year and a half, along with the detailed FAA report and the voice tapes, which had been placed on that corner table waiting for the CIA. No one ever came and got them. When I was leaving for retirement in August 1988, one of the branch managers, in a hurry to get me out, packed everything that was hanging on the walls and sitting in the office, put it in boxes, and shipped it to my house. I've had this data and the video in my possession ever since.

Now, more than twenty years later, it's become very clear to me that most people, including FAA controllers, really aren't familiar with how the FAA radar system works and why all aircraft traveling through our airspace are not caught on radar or displayed on the controllers' PVD. The system and organization of the FAA are not configured to identify and track these aircraft types. In short, current FAA equipment will not paint a "spaceship" unless the aircraft has slowed to a speed similar to current aircraft.

The reasons are simple: The UFOs appear to have no transponder; they are often too big for the automation system to be considered an aircraft, so the radar thinks they're weather (radar readings with an unrecognizable signature are often automatically sent out through a second system as weather); or they're too fast for the radar to get a hit on before they're out of range. If something is hovering, as it was at O'Hare Airport in 2006, it often doesn't show up, or if it did it would be a small dot and FAA controllers would not give it much concern.

During the playback of the 1986 event I clearly observed a primary radar target in the position reported by the Japanese pilot. But the radar signals were intermittent because the UFO was painted as an extremely large primary target and so the FAA computer system treated the UFO radar return as weather. Regardless, the target could be seen near the 747 off and on for thirty-one minutes.

So we have a problem. Because of these radar deficiencies, when pilots report seeing an unusual object, the FAA will not investigate unless the object can be identified by an airborne pilot, and instead the FAA will offer a host of weak explanations. If the FAA cannot identify the object within FAA terminology, then it doesn't exist. Another cliché we sometimes used: For every problem there is a solution. The FAA seems to believe that the converse is also true: If there is no solution, there is no problem.

The Alaska UFO investigation is a case in point. The final FAA report concluded that the radar returns from Anchorage were simply a "split image" due to a malfunction in the radar equipment, which showed occasional second blips that had been mistaken for the UFO. Thus the FAA would not confirm that the incident took place.

Yet all three controllers engaged with the pilots during the extended sighting filed statements that contradict this finding. "Several times I had single primary returns where JL1628 reported traffic," wrote one. "I observed data on the radar that coincided with information that the pilot of JL1628 reported," stated another.

The FAA spokesman at the time, Paul Steucke, said it was just a "coincidence" that the split image happened to fall at the right distance and the same side of the aircraft where the object was reported visually by the pilot. And the final report simply outright ignored the three visual sightings with all their details and drawings, as if the event really had never happened. Remember, no one flying an aircraft can see a split image.

So, who are you going to believe, your lying eyes or the government?

CHAPTER 23

Government Cover-up: Policy or Myth?

The CIA's directive that "this event never happened," as reported by former FAA official John Callahan, may be familiar to those who have read statements from American military witnesses to UFO events. Many have been told more or less the same thing by their superior officers: Do not speak to anyone about the incident that you just experienced. In later years, some say they still cannot speak publicly because they're bound by security oaths, and no doubt there are many others who, out of fear of breaking such oaths, have not even hinted of their involvement in a UFO event while in the military. But a number of fearless men and women have, years later, spoken out in spite of orders or oaths, without repercussions.

This repeated demand for silence, coupled with the overzealous classification of government documents and the furtive misidentifications issued by Project Blue Book, and later the FAA, has led to much speculation about whether government agencies are involved in some kind of cover-up—a widespread, carefully orchestrated policy, hidden from almost everyone, to keep secret "the truth" about UFOs. While publicly ignoring and avoiding the UFO issue, underneath the surface and unbeknownst even to those issuing the orders muzzling subordinates, a small yet powerful core group is actively hiding explosive knowledge, such as the extraterrestrial origin of at least some UFOs. At least this is what many—even conservative—analysts have come to believe.

As far-fetched as it sounds, this radical supposition cannot be dismissed out of hand. Documents prove that the UFO phenomenon became a concern to the Air Force, the CIA, and the FBI as long ago as the late 1940s, thereby giving U.S. authorities ample time to collect the best data and study physical evidence. Obviously the military would have been extremely interested in the technological capabilities demonstrated by these objects, if they could ever get access to them. We must consider the possibility that enough concise data—even physical material retrieved from crashed UFOs—could have been obtained and studied in secret. If our government officials were hungry to discover some of the keys to these exotic new technologies, or thought we were on the verge of unearthing a new physics, something from another space-time perhaps, these discoveries could give America unimaginable new capabilities.

Of course, such a study would have been daunting and could take decades. No matter how intense, scientists might still not be able to figure out very much about the workings or origins of UFOs, given the sophisticated, perhaps undecipherable technological systems, so remarkable that they seem almost like magic to us. The analogy has been made to a group of cavemen suddenly coming into possession of a television set, before even understanding the fundamental concepts of electricity or radio waves. Of course, this is pure speculation. But even if our covert scientists made very little progress on understanding what we had, it's not a stretch to imagine that those in charge would have been extremely careful to keep such revolutionary information away from any "enemy" countries or rogue nations, including the Soviet Union during the Cold War. They would have been mindful of any future economic benefits that could result from these exotic technologies as well, and would likely want to ensure that U.S. corporations would be the exclusive beneficiaries of any breakthroughs.

As discussed previously, some official documents of the forties and fifties clearly show that, having eliminated the option of the phenomenon being some new manifestation within the natural world, a number of highly placed officials *did* take the position that UFOs were interplanetary. An inclination to withhold from the public information about something so unthinkable is conceivable given its potentially vast implications. Perhaps those in possession of the secret just wanted to put off its release until more was learned, but that day never came. Also, reflecting back to Nick Pope's "threat = capability + intent" equation, there would have been much concern about inherent dangers. A rational governmental response would have been to understand and control the situation as

much as possible before acknowledging anything about unidentified flying objects to anyone else, and to keep that explosive information highly
classified. Our government would not have wanted to risk mass hysteria.

Obviously, we don't know with any certainty whether or not such a
secret research program exists, although there have been hints and suggestions, usually from reports of individuals claiming indirect knowledge,
that keep the question alive. It is raised repeatedly by those curious about
UFOs, many of whom regard it as an issue of major, compelling importance. However, the alternative notion is much easier to accept: that the
United States is as baffled as anyone else about this mystery, and just
as helpless in confronting the unpredictable phenomenon as any other
country. The world's superpower simply shrugs its shoulders and looks
away, as if there is nothing to be done, focusing on more urgent matters
confronting human beings than the sporadic appearance of something
odd in the sky.

The fact is, even if we eventually learn that a secret research group
has been operating, the state (meaning the government, military, and
scientific structures creating our society) is undoubtedly not privy to
this intimate information about UFOs. Any behind-the-scenes endeavor
would have to be *so* exclusive, so entirely covert, that in effect its existence would make no difference to our government or country, to the
people who know nothing about it, which is essentially everyone. In this
sense, it's unimportant to the business at hand: establishing a U.S. agency
so that an open, worldwide investigation can take place.

Nonetheless, even though the question of a cover-up is really a side
issue, and will continue to be as long as such a program—if it exists at
all—remains deeply buried, it remains a focus for the interested public,
hotly debated and often explored in television documentaries. In interviews about UFOs, it is usually one of the primary questions asked.

When I first became involved with the subject of UFOs, I sought out
reliable, trusted sources as any responsible investigative journalist would,
attempting to find out what our government actually knows about UFOs.
The process took many years, requiring great care and discernment,
and eventually sources began seeking me out as well. Whether or not I
choose to take any person seriously comes down to personal judgment,
which for me is based on meeting the person whenever possible, talking
at length, knowing them over time, learning about their background,

checking the accuracy of facts they've reported, and understanding their motivations. Also, I always look for corroboration.

When probing the question of a possible secret government research program into UFOs, or anything, for that matter, that is highly sensitive, the sources will rarely go on the record, for obvious reasons. Their accounts are also extremely hard to verify, because even if they provide names of others involved, these people will deny any knowledge of such a program. Alarm bells could be set off by an attempt to locate these individuals, so sometimes I have been asked not to do so. This type of information, therefore, as exciting as it may be, has to be relegated to what reporters call "deep background." It may help inform how one views the issue, but not centrally. It can nudge us in a certain direction, or inspire future inquiries. It's all very intriguing, but always just out of reach.

I am willing to take such sensitive information seriously when two or more credible, qualified sources report the same thing *independently of each other*—for example, when men from different branches of government who don't know each other, with years separating their statements, provide essentially the same reports. And concerning the question of a secret government research program on UFOs, this has occurred. A number of reliable sources have told me about their conversations with high-level military contacts who say they are aware of a deeply hidden program for UFO research, one which is so closely guarded that even people at the highest levels of the military are denied access to it. Some of these independent accounts include names and specific details. Much case evidence over the years has also pointed to the plausibility of this kind of program, although it can't authoritatively be determined one way or the other.

Some of the anonymous sources I refer to include mainstream scientists, all Ph.D.s with impressive, lengthy resumes, some of whom have worked for the CIA or other intelligence agencies—an astrophysicist, a physicist, an astronomer, among others—and a NASA aerospace engineer. One military source, Commander Will Miller, U.S. Navy (Ret.), has gone on the record while keeping certain specifics confidential. He agreed to reply to a series of questions I presented to him in late 2009 about the question of government secrecy.

Although still very active, Miller, who now lives in Florida, retired from active duty in 1994, the same year he was awarded the Department of Defense Meritorious Service Medal. As a naval officer and decorated Vietnam combat veteran, he had his own sighting from a Navy vessel while serving near Vietnam. He later became a senior Department of Defense command center operations action officer, a senior intelligence analyst, and a program manager for DoD future operations programs such as WWIII planning, nonlethal weapons systems, and future space systems. He was an advisor to U.S. Space Command and U.S. Southern Command and its international counterdrug operations, Joint Interagency Task Force East. As an expert in special contingency operations, Miller held a Top Secret clearance with Sensitive Compartmented Information (SCI) access, meaning he had access to sensitive information whose handling is restricted one step further than the Top Secret classification, including that which is related to topics and programs not publicly acknowledged.

While an officer on active duty throughout the 1980s, Miller did not hide his interest in UFOs. "I was simply a concerned officer who studied the subject, looked at the facts, and talked to people in the military," he says. "People with personal knowledge would seek me out because they knew I had an interest. I've done this for a long time."

By 1989, Miller had become acutely aware that high-ranking military officers were not properly informed about the UFO phenomenon, and he became concerned, like the COMETA authors, about possible national security issues arising, not from the UFOs themselves, but from a lack of preparedness. He believes that we must assume UFOs have the same right of self-defense to hostile intent or hostile acts as we accord our own military forces. Fortunately, these rights have not been acted upon by the UFOs, as far as we know, when attacked. "Only a small fraction have demonstrated even a remote semblance of hostility, and that was only with severe provocation, usually an attack by military aircraft," he says. "If the entire body of data were examined, the obvious conclusion would be that UFOs are not hostile. That is precisely what the U.S. military declared after many years of UFO study: that UFOs pose no threat to the national security of the United States."

After he retired from the U.S. Navy, Miller began taking steps to set up a series of information briefings that culminated in meetings in 1997 with Vice Admiral Thomas R. Wilson, vice director for intelligence on the Joint Staff, and in 1998 with Lieutenant General Patrick M. Hughes, director of the Defense Intelligence Agency (DIA). (Wilson later became director of the DIA and Hughes, the corporate vice president for intelligence and counterterrorism at the Department of Homeland Security.) Miller has provided me with a confidential, detailed account of these meetings and those leading up to them, including attendees, preparatory briefings, topics discussed, and reactions from attendees.

He explains that he raised two national security concerns at his briefings: the risk of uninformed human aggression toward UFOs, leading to a possible disaster, and the government's disregard for public concern about UFOs and its refusal to provide honest answers to legitimate questions. Miller feels strongly that unnecessary secrecy threatens the public's sense of personal security while eroding trust in the government institutions mandated to inform and protect U.S. citizens. "The officials have universally received these briefings with the same serious consideration as briefings on any other national security issue," he says.

I first contacted Commander Miller ten years ago, in 1999, through the introduction of a mutual colleague. I was repeatedly struck by the similarity of his conclusions and approach to those expressed by retired French military officers in the COMETA Report, communicated to me before Miller had any idea of its contents. He and the French officers had all been through a similar process to arrive at these positions, but within two different militaries. They were all meticulously careful about what they said, suggesting that they knew more than they could reveal. Of course, Miller has never had the strength-in-numbers of the French group—his is a lone voice in a vast wilderness, by contrast, and a particularly courageous one given the risks to his reputation through association with the UFO subject.

I sent him a confidential copy of the translated COMETA Report while writing my first UFO story for the *Boston Globe*. I then spent many months in substantive telephone interviews with him, and we met in person a year later. Over time, I came to know and trust him as a person of integrity, clarity, and devotion to his country, and have regularly consulted with him about issues involving UFOs and the military. Well connected at high levels within the impenetrable military and intelligence world, Miller is a true "insider" of the highest order. He is one of the few who has persistently taken his concern about UFOs to authorities above him, and has spent many years assessing the official relationship to the phenomenon through his access to American generals, admirals, NSA contacts, and other sources of sensitive information.

"The military officers I talked with were extremely interested in getting factual information on the UFO subject, since even at the flag-officer level, they were unable to get that information through normal military intelligence channels," Miller told me. Throughout the years, as he continued to speak with his contacts, he became more and more convinced of the existence of a well-concealed, "need-to-know" UFO program, based on statements that he says confirm this fact, made by military personnel attending his Pentagon briefings and others.

I asked him in late 2009 about his overall assessment. He replied in an e-mail:

1. It is a fact that there are those in high places in the government who have an interest in this subject (in many cases it is because they or a member of their immediate family has had a sighting or personal experience with the phenomenon).
2. When the American people say the government is in the middle of a massive cover-up, in most cases that is absolutely NOT the case; those people in positions where you would say "they've got to know" absolutely don't.
3. I remain firmly convinced that many military and civilian personnel at the highest levels of various agencies, departments and organizations are purposefully kept in the dark so that those leaders may plausibly and honestly deny knowledge of the subject.

I asked Miller to elaborate further on who is keeping whom in the dark:

The "control group" cannot allow any information on their closely held UFO research to be accessed by anyone outside of those specially cleared for that Unacknowledged Special Access Program (USAP). Neither Joint Chiefs of Staff Intelligence nor the director of DIA himself could get ANY information on the subject; this is a fact. Yet I know that sources within multiple organizations maintain such information. Leadership remains "protected" from such knowledge. As far as I am concerned, the question is answered.

He added further comments on the issue of secrecy:

> To the best of my knowledge, members of the Joint Staff in general are only aware of UFOs and any related secrecy issues from what they read and watch on TV. In fact, there are no secrecy issues related to UFOs since the consensus is that they have not been proven to exist and therefore do not hold a place in the list of secrecy topics about which Joint Staff members are forbidden to speak. That said, however, if a person were to encounter documents or other information related to the subject of UFOs that were classified, then that person would be bound not to discuss that classified information.
>
> The phenomenon is ignored as if it was an unproven myth despite the existence of classified information about it. I know for a fact that such information resides within several "three-letter agencies." That is no surprise, since multiple agencies in the past have tracked these objects, received reports on these objects, and created reports related to military and/or civilian encounters with these objects and/or their effects. Especially where surveillance and detection systems are concerned, a reasonable person might assume that agencies tasked with detecting and monitoring air, space, and sea via various technical surveillance systems would periodically detect these UFOs/crafts or have reports of such sent to them, which they would then disseminate to appropriate authorities/end users with the need to know.

Would it be possible to keep something like this secret? CDR Miller referenced the possibility of an Unacknowledged Special Access Program (USAP) as one potential location for a group controlling access to UFO information. USAPs are one of the known mechanisms in place within the Department of Defense for controlling sensitive information without public knowledge of its existence. An investigative report by Bill Sweetman in *Jane's International Defense Review* sheds tremendous light on the extent to which the DoD is capable of keeping secrets. These "black projects" within the DoD, officially called Special Access Programs (SAPs), are structured so that those involved in one component do not know what is going on in another, preventing knowledge of the bigger picture. Buried even deeper is the USAP referred to by Miller, a black program so sensitive that the fact of its existence is a "core secret," defined in U.S. Air Force regulations as "any item, progress, strategy or element of information, the compromise of which would result in unrecoverable failure." This means that all participants are required to deny the very existence of the program if confronted, since even "no comment" is considered a confirmation.

Cover for these projects is supported by "the dissemination of plausible but false data, or disinformation." Often, the false information accompanies some truth, so that the two are indistinguishable and the truth is thereby discredited. "Presented with a wall of denial, and with no way to tell the difference between deliberate and fortuitous disinformation, most of the media has abandoned any serious attempts to investigate classified programs," writes Sweetman. Perhaps, as has been revealed on occasion throughout the decades, some of the leaked "official" documents and shadowy characters with wild claims, emerging from the deep, dark intelligence world, could be part of an official disinformation program, protecting the USAPs exclusive ownership of the truth by confusing those getting closer to it. We simply don't know.

In 2008, I acquired an extremely interesting document from the UK, quietly released to a researcher through a FOI request. It comes close to verifying the existence of such a secret group in America—the *only* legitimate, confirmed government document to do that, to my knowledge. It so happens that it was written in 1993, during Nick Pope's tour of duty at the Ministry of Defence's "UFO desk," and that he played a role in its conception and execution. Titled "Unidentified Aerial Phenomena Study" and running just over one page, the document is a proposal for a study (which was approved and became Project Condign, described in chapter 17). Initiated by the Defence Intelligence Staff (DIS), it needed the approval of Pope's department. Written by his counterpart in the DIS, it was addressed to Pope's superior "Sec(AS)2," the Air Staff deputy director, and classified "Secret UK Eyes A."

The key section is paragraph 2, with two parts redacted and replaced with rows of the letter X:

2.   I am aware, from intelligence sources, that xxxxx believes
that such phenomena exist and has a small team studying them. I
am 'also aware that an informal group exists in the xxx
xxxxxxx community and it is possible that this reflects a
more formal organisation.

After careful consideration based on deductive reasoning, I offer the following analysis.

Before spending resources on any study, the first thing a government agency would do is check with its allies and find out what they may
already have learned about the subject being considered. It is reasonable
to assume that UK intelligence would check with its number one ally, the
United States, through its own sources in the intelligence community,
since intelligence officers, like the one who wrote this proposal, work
directly with their counterparts within other countries. Secondly, it is also
logical to assume that UK intelligence would be interested in the work of
any major countries of concern, important players that may be adversar-
ies and thus are monitored on a regular basis. In this case, that country
would be Russia.

The next step is to go back to the document and see if these countries
would physically fit in the spaces with the X's. The number of X's used in
the redaction process does *not* necessarily correspond with the number
of missing letters. Therefore, when seeing what fits, one has to look at
the amount of space, not the number of X's. It just so happens that the
word "Russia" fits in the first line, and the words "US intelligence" fit perfectly in the second line in the two spaces provided, when measuring the
length of the words in relation to the space, and also in keeping the spac-
ing between the words consistent within each line. Replacing the X's, the
document would then read (emphasis added):

> I am aware, from intelligence sources, that *Russia* believes that
> such phenomena exist and has a small team studying them. I
> am also aware that an informal group exists in the *US intelligence*
> community and it is possible that this reflects a more formal
> organization.

The meaning and implications of these two lines, especially the second one, are well worth considering. Line one is actually not surpris-

ing, since a great deal is known about longstanding Russian research and military interest in the UFO phenomenon. In line two, the statement "I am aware" means that the writer is stating a fact: the informal group exists. An "informal group" is one which provides nothing in writing and leaves no records behind, one that escapes oversight by House or Senate committees, and might be set up this way because its work runs counter to established policy. It could be part of a SAP. As defined by Nick Pope, "an informal group would be a loose network of individuals, perhaps in a number of different agencies, coming together to discuss a particular issue, but without formal terms of reference."

The second half of this sentence begins with "it is possible"—unlike "I am aware," this phrase is *not* stating a fact, but only a possibility. This too is actually quite revealing. One must question why this intelligence officer could not get more information about the nature of this group from his closest ally. He has not been told much at all about the nature of the "informal group" and was not able to determine whether this reflected any more of a "formal" structure, something properly constituted. This attests to the highly secret, deeply buried nature of the informal group.

If indeed this interpretation is correct, and I have every reason to believe that it is, this document references a secret group within the U.S. intelligence world actively studying UFOs. It is a much more important piece of paper than any new case reports released recently by the MoD, which have received all the attention. The public position of the U.S. government is that they haven't investigated UFOs since 1970, when Project Blue Book was closed down. But this British document—the provenance of which is beyond dispute—potentially blows this claim out of the water. According to this analysis, the United States *is* studying UFOs. But by doing so in the way revealed by the UK, the program is operating behind the backs not only of the public and the media, but also of Congress, the Senate, and the president. However, this is in no way "proof" or definitive since we'll never receive proper confirmation as to the missing words, which remain classified.

I approached Nick Pope hoping to receive some clues, some hidden message. But he is too much of a pro ever to be caught off guard. He acknowledges that he helped his DIS colleague with the drafting of the proposal for the UFO study, and can recall which two countries were redacted from the document. I asked him about my assessment of the two missing words, and whether he could respond in any way to it. "No comment" was his reply.

This material, though intriguingly suggestive, is in no way definitive. Taking a step back, we must reflect once again on what we actually *know*, in order to move forward. Caution, or even understatement, must be the name of the game when dealing with the unaccepted subject of UFOs. The reality of what we *do* know is extraordinary enough.

For many, the process of discernment is not easy. Conspiracy theorists and the television media have fueled an intricate, rumor-based mythology around the idea of a cover-up, leading some to write off the whole subject of UFOs as inane science fiction, and others to swallow every morsel offered. Those in the middle have no way of sorting out the valid information from the fanciful, which get all mixed into one big pot of unhealthy soup. (This is essentially *self-made* disinformation, and no secret agents are needed to disseminate it since the media and large swaths of the so-called UFO community take care of that themselves.) But behind all the extreme reactions is the actual fact that the state doesn't seem to want us to know UFOs exist. Since we know they *do* exist, we have to assume that the government knows that, too. If so, why is it hiding this, and *what* is it hiding? People are desperate for answers, and very frustrated, and they have understandably come to deeply mistrust our government on this issue.

Some take advantage of this situation. So-called whistleblowers at varying levels of psychological health and mental clarity regularly jump into the pot—people who have no relationship to the credible sources I referred to earlier—claiming direct knowledge of some aspect of a sinister government cover-up. Undiscriminating UFO groups have made them or their spokespeople into heroes and trotted them out at press conferences, offering them up like sacrificial lambs to be promptly ridiculed by the few media that bother to take note. And in many of these clearly unfounded cases, the ridicule is well deserved. Others market themselves as scholars or activists, making baseless accusations and claims about government misdeeds regarding UFOs, based on rumor rather than record. These extremists only serve to muddy the waters and compound the public relations nightmare that UFOs already face within public discourse. Sadly, this is the only kind of UFO information that so many Americans have been exposed to.

Putting the hype aside, serious investigators and retired officials make the legitimate point that the known facts alone, such as those raised so far in this book, *do* lead to perplexing, unanswered questions about U.S. government secrecy. In 1999, the French COMETA group chastised the United States for what it calls an "impressive repressive arsenal" of tactics protecting UFO information, including a policy of disinformation and military regulations prohibiting public disclosure of sightings. Air Force Regulation 200–2, "Unidentified Flying Objects Reporting," for example, prohibits the release to the public and the media of any data about "those objects which are not explainable." An even more restrictive procedure is outlined in the Joint Army Navy Air Force Publication 146, which threatens to prosecute anyone under its jurisdiction—including pilots, civilian agencies, merchant marine captains, and even some fishing vessels—for disclosing reports of those sightings relevant to U.S. security. Fortunately, I am not aware of any cases in which such extreme actions were taken.

But we do know for sure, as shown by the Bolender memo and government files released through the FOIA, that the U.S. government has had *some* level of involvement in UFO investigations since the close of Project Blue Book, despite claims to the contrary. Nevertheless, officials are usually irrationally unresponsive to unfolding UFO events, as they were during the Hudson Valley sightings in the 1980s, and provide ridiculous and false explanations when pressed.

We also know that UFO documents have been previously classified by government agencies, as shown by their later release through the Freedom of Information Act, and that some information still remains so. National Security Agency UFO files were released in 1997, following a lawsuit years earlier, but they were so heavily redacted (the NSA stated all deletions had to do with protecting sensitive sources and methods) that they were virtually useless. In response to FOIA requests, agencies have initially denied having documents on file which turn up later somewhere else, or are found in a second search. Researchers have discovered that in many UFO cases for which official reports were filed at the time, none can be found later when looking in the logical places. And as also stated in the Bolender memo, UFO reports affecting national security were to be filed outside the Blue Book system. Where *are* these files, and why can't they all be released?

Over the years, even senior government officials have made an effort to access hidden UFO evidence. Senator Barry Goldwater attempted to penetrate the vaults at Wright-Patterson Air Force Base, the home of Project Blue Book, during the UFO "golden age" of the 1960s, and described his efforts in a series of letters he wrote in response to inquiries years later. Goldwater, a licensed pilot and retired major general in the U.S. Air Force Reserve, had studied reputable pilot reports and had a longtime interest in the subject. He was convinced that a secret UFO program *did* exist. "About ten or twelve years ago I made an effort to find out what was in the building at Wright-Patterson Air Force Base where the information is stored that has been collected by the Air Force, and I was understandably denied this request. It is still classified above Top Secret," he wrote in a 1975 letter.

In a 1981 letter to a researcher, Goldwater said that, regarding this effort, "I have had one long string of denials from chief after chief, so I have given up . . . this thing has gotten so highly classified, even though I will admit there is a lot of it that has been released, it is just impossible to get anything on it." And in 1983 he wrote: "I have no idea of who controls the flow of 'need-to-know' because, frankly, I was told in such an emphatic way that it was none of my business that I've never tried to make it my business since."

Finally, when asked during a 1994 radio interview, Senator Goldwater said: "I think the government *does* know. I can't back that up, but I think that at Wright-Patterson field, if you could get into certain places, you'd find out what the Air Force and the government knows about UFOs . . . I called Curtis LeMay and I said, 'General, I know we have a room at Wright-Patterson where you put all this secret stuff. Could I go in there?' I've never heard him get mad, but he got madder than hell at me, cussed me out, and said, 'Don't ever ask me that question again!' "

A year later, in 1995, New Mexico congressman Steven Schiff announced the results of a General Accounting Office (GAO) investigation, which he initiated on behalf of his constituents, attempting to access records related to events surrounding a mysterious crash in 1947, near Roswell, New Mexico, which has become famous due to the popular belief that what came down was a flying saucer. "The GAO report states that the outgoing messages from Roswell Army Air Field (RAAF) for this period of time were destroyed without proper authority," Schiff explained in his press release. "It is my understanding that these outgoing messages were permanent records, which should never have been destroyed. The GAO could not identify who destroyed the messages, or why." The Air Force had claimed for nearly half a century that the crashed object was a weather balloon. In 1994, while Schiff was waiting for results from the GAO, it retracted that statement and announced that the crash debris actually came from a then-classified device to detect evidence of possible Soviet nuclear testing. Naturally, that delayed explanation raised enough new questions to keep the Roswell controversy going, one that includes a volume of compelling witness testimony contradicting the Air Force position. The unsuccessful efforts of both Goldwater and Schiff to obtain information through official channels do not *prove* a cover-up of knowledge about what UFOs are, as so many would like to believe, but they do reveal how difficult it is to acquire definitive information about UFOs from the U.S. government.

In fact, each component used to argue that excessive government secrecy shows there is an official cover-up of knowledge about UFOs could have a host of possible alternative explanations. We know that the FOIA does not work efficiently, and that the complicated bureaucracy involved with record keeping is overwhelmed and not well organized. UFOs might logically be on the bottom of the list of priorities. And where are all those who would have worked on this deep, black program—hundreds or thousands of specialists, or their surviving family members? Certainly at least a few would feel the moral imperative to share knowledge or discoveries about UFOs with the rest of humanity, and would take the risk of doing so, perhaps even seeking shelter in whistleblower protection programs. And yet there have been, so far as we know, no deathbed confessions or willed documents from any of these government scientists, nor have any wives revealed the truth about a Special Access Program on UFOs. Not even one. And finally, we have not seen the results of any truly fantastic back-engineered military technology that might have resulted from captured UFOs, despite rumors to the contrary.

Directives to military and government employees instructing them to keep sensitive matters quiet are standard operating procedure for a range of issues and a range of purposes. The sudden appearance of an unknown object creating havoc for Air Force pilots at sensitive air bases would not be something any military authorities would wish to make public, especially during the Cold War. If the military was unable to identify that something, it seems even more logical that the event would be kept under wraps. But this does *not* mean that there is a cover-up specifically of UFOs, or that we ever learned what the nature of these unknowns may have been. A host of national security concerns can compel government secrecy, and the military always prefers to err on the side of excessive secrecy rather than the opposite.

Returning to the easier analysis, perhaps the sensitive research projects hidden within the U.S. government avoid dealing with UFOs simply because even our most specialized intelligence officials actually don't know much about them and can see that there is nothing to be done one way or the other. The objects haven't caused us harm and there are many other, more immediately dangerous and pressing issues to be addressed, involving human survival both economical and environmental. This would mean that the only cover-up in place is that which conceals any recognition that UFOs exist, and involves nothing more than that.

And this nonacknowledgment has its own logic. It makes sense that the authorities would have no motivation to announce publicly that there are apparently all-powerful unknown machines flying without restriction in our skies and beyond our control. Would our government want to acknowledge its own impotence in the face of something unidentified yet well-documented? Some authorities may worry about public panic, whether we know what they are or not. Even if the U.S. government acknowledged the presence of an unexplained phenomenon, the extraterrestrial hypothesis would become part of the debate, and if the thinking became that these likely *are* vehicles or drones from somewhere else, it would appear that they have complete power over us. What official body would want to unload such a bombshell in an already unstable world?

On the other hand, it is important to remember that the Belgian Air Force did just this in 1990, and other countries have done so, as well, in relationship to specific events, and no dire popular upheavals or waves of fear have disturbed these societies. Instead, people continued their regular lives with much less need than we find here in America to create alternative explanations or conspiracy theories in order to satisfy their natural human curiosity. Nonetheless, in this huge, multicultural country that sees itself as a planetary leader on many fronts, opening that door through any kind of organized official statement seems to remain entirely unappealing.

However, such government reticence must and *can* be overcome, or at least outflanked, according to former governor Fife Symington of Arizona, who has had unique experience—to say the least—on both sides of this complicated fence, leading to his current stance on the issue. Beginning with the landmark press conference of 2007, he and others from around the world have formed a united platform seeking a new approach. The citizens of the world, including Americans, are ready to move forward.

CHAPTER 24

Governor Fife Symington and Movement Toward Change

O n March 13, 1997, a decade after the Hudson Valley UFO wave had quieted down, multiple triangular and V-shaped UFOs made a series of brazen new appearances, this time over the western United States.

It was a pleasant spring evening in Arizona, clear and still, and countless families were outside in larger than usual numbers gazing at the sky, because Comet Hale-Bopp was to be visible that night. Instead, beginning at about 8:00 p.m., they were provided with an even more astounding aerial spectacle: a series of massive, eerily silent craft gliding overhead like nothing they had ever seen before. One central object moved from the north, southeast across the state, traveling about 200 miles from Paulden to Tuscon, passing near Phoenix and surrounding communities. It was on display between 8:15 and 9:30 p.m. Many hundreds—more likely thousands—saw it.

Police department phone lines were jammed and the local air force installation, Luke Air Force Base, was overwhelmed with calls. Reports of sightings from around the state flooded the lines at the National UFO Reporting Center (NUFORC)—the well-known repository for UFO reports cited in the FAA manual—based in Seattle, Washington. Even so, air traffic controllers apparently did not register the strange objects on radar.

Although descriptions of the array of lights differed, one overriding characteristic prevailed: the craft was massive; it was a solid object, not merely lights; and it often appeared to be very low in the sky, blocking out the stars behind it. A younger witness said he could clearly see the underside of the craft, and thought if he had thrown a stone, he could have hit it. According to eyewitness reports in the NUFORC files, which received its first report at 6:55 p.m. from Henderson, Nevada, one group of three said it blocked out most of the sky, while another family of five described looking out the automobile windows while driving at eighty miles an hour and observing the incredibly huge craft passing above their car. It was the size of multiple football fields and up to a mile long, many said. A little league game had to stop as the massive object passed over the heads of moms, dads, kids, and coaches staring in disbelief. Some people described its color as a dark gun-metal gray, and many people were awestruck by the silence of the object, given its size, especially when watching it take off in the blink of an eye.

It was difficult to determine how many objects were present, because reports varied in terms of the number of lights, colors of lights, and movements. The speed of the craft, or crafts, varied from motionless to speeding away in an instant. Calls came rapidly into NUFORC from many communities at different locations, suggesting the likelihood that multiple objects were cruising overhead, some perhaps moving rapidly between locations. It took many months for the civilian investigators who took on the case to compile all the reports, map the trajectories, and determine that indeed several objects had been seen.

Once again, as in the Hudson Valley wave, no government officials were dispatched to investigate or respond to questions from alarmed and awestruck citizens. To put it bluntly, in 1997 the federal government failed to react to the presence of something huge and unknown invading restricted airspace over a capital city in the United States of America.

Phoenix city councilwoman Frances Emma Barwood, responding to pressure from journalists and her constituents, was the only elected official to launch a public investigation. But she said that she too received no information from any level of government. Barwood says she spoke with over seven hundred witnesses who called her office, including police officers, pilots, and former military personnel, all providing very similar descriptions of the objects. Still, government officials seemed uninterested. "They never interviewed even one witness," Barwood told me in a conversation a few years ago. "How could they possibly not know about these huge craft flying low over major population centers? That's inconceivable, but it's also frightening."

Due to her willingness to respond to public concerns about the incident, Barwood was ruthlessly ridiculed by much of the Phoenix media, including a well-known cartoonist in Arizona's leading newspaper, and she also suffered from disparaging comments by male political figures. "What happened to me was a lesson for other elected officials," she told me. "If you talk about this, you will get ridiculed, chastised, pummeled with everything you can imagine, and eventually lose credibility."

Minimal coverage was provided at the time of the incident by the media, even in Phoenix, with a few local papers and news stations making note but not following up. Three months later, on June 18, that all changed when *USA Today* brought the case into the national spotlight with a front-page story. It was further catapulted onto the network evening news when the sightings were covered, although very minimally, by ABC and NBC, and became known as the "Phoenix Lights."

By the time the *USA Today* story broke, pressure had been mounting within the state of Arizona and public reaction was intensified by this new level of national media attention. Frustrated citizens wanted answers. The next day, on June 19, Republican Governor Fife Symington announced on morning television that he was ordering a full investigation and would make "all the necessary inquiries. We're going to get to the bottom of this. We're going to find out if it was a UFO."

Later that afternoon, he called a press conference, telling people that he would reveal the source behind the Phoenix Lights. With an excited media covering it live, and citizens glued to their sets awaiting the news, Symington shocked some, angered others, and amused many more when he presented his "explanation." His six-foot-four Chief of Staff, Jay Heiler, handcuffed and wearing an alien outfit featuring a large, gray rubber mask with huge black eyes that fitted over his entire head, was escorted to the podium by public safety police officers. The governor presented the Halloween-costumed extraterrestrial as the "guilty party." While laughter filled the room, he joked that "this just goes to show that you guys are entirely too serious," and the mask was removed before the cameras.

Symington also announced that he'd made inquiries with the commander at Luke Air Force Base, the general in charge of the National Guard, and the head of the Department of Public Safety, but had not learned anything at that point. This important statement was overshadowed by reactions to what he now calls his "spoof" press conference.

As one might expect, Councilwoman Barwood's office was bombarded by calls from outraged people, and the governor received his share of complaints as well. Unable to get anywhere on her own, Barwood approached Arizona's Senator John McCain and asked him to conduct an investigation. McCain asked the Department of the Air Force to investigate, and as he explained in an October 1997 letter to a constituent, "The Air Force informed my office in July the Department no longer conducts investigations into UFOs." McCain went on to explain that local military installations, however, did "make an effort to resolve the issue" by checking records from that night, and he was informed that the National Guard had dropped high-intensity magnesium flares southwest of Phoenix between 9:30 and 10:00 p.m., which could be seen for 150 miles.

Indeed, television news and documentaries about the Phoenix Lights repeatedly feature a video taken at around 10:00 p.m. by an amateur photographer, as if it represented actual footage of the UFO. The now-infamous video has been subjected to detailed analysis by at least two qualified professionals, and both determined that the brilliant lights shown hanging in a row over the mountain ridge and then dropping out of sight, were, in fact, flares. Since the amateur video was taken at 10:00 p.m., at the same time that the National Guard states it was dropping LUU2 flares as part of a training exercise known as "Operation Snowbird," and the photo analysis confirms that the lights in this film *were* almost certainly flares, the questionable later video is not the evidence many people had wished for. This fact seems to be overlooked by the media, hungry for something visual when they report on the story.

The time of the flare drops is extremely important. The most widely viewed sightings of unidentified objects across Arizona that evening began at approximately *8:15*, although some objects were viewed earlier in daylight. Clearly, the UFO flyovers were a separate event occurring independently of the later flares.

It is interesting that in his letter, Senator McCain, a longtime friend of Governor Symington, informed his inquiring constituent that he was still exploring other possible explanations. In a 2000 press conference, McCain acknowledged that an incident in which mysterious lights were seen over Arizona had actually occurred. "That has never been fully explained," he said, "but I have to tell you that I do not have any evidence whatsoever of aliens or UFOs." That same year, a class action suit was filed in U.S. District Court in Phoenix by witnesses demanding an explanation from the federal government. In response to a court-ordered request for a search for this information, the Department of Defense maintained that it could not find any information about the triangular objects. It provided details of this search process to U.S. District Court judge Stephen M. McNamee. On March 30, 2000, three years after the sightings, McNamee concluded that "a reasonable search was conducted," even though no information was obtained, and he dismissed the case.

We have no way of gauging how thorough this search really was. And the claim of the DoD seems open to question, especially in light of a prior British inquiry about the triangular craft observed over the Royal Air Force base at Cosford. As reported by Nick Pope, this object was seen by over a hundred witnesses in England in 1993, including police officers and military personnel. At the time, the MoD sent a discreet letter to the U.S. Embassy that was "disseminated to all 'interested Agencies' in the U.S." to find out whether the Cosford object could have been attributable to some secret U.S. prototype such as the Aurora. In response, the American officials said that they had been having their *own* sightings of these large, triangular-shaped UFOs and wanted to know if the RAF might itself have such a craft! This remarkable reply amounts to an acknowledgment by American officials—which probably they did not expect would be made public—that in 1993 they were aware of the existence of unexplained objects operating over the U.S.A. with the extraordinary capabilities attributed to the Cosford UFOs. Perhaps they were alluding to the Hudson Valley wave of the 1980s, although other sightings had occurred since. Importantly, these officials recognized the similarity between the Cosford object and the ones seen here, and were sufficiently perplexed to express their hope that the American UFOs may have been secret British aircraft flying without authorization, an extremely unlikely proposition given our close relationship with the UK. After this exchange, the British MoD laid the Cosford incident to rest. "None of the usual explanations put forward to explain UFO sightings seem applicable," the MoD stated. The evidence showed that "*an unidentified object (or objects) of unknown origin* was operating over the UK" (emphasis added). U.S. officials had inadvertently acknowledged, privately and secretly, of course, that this was true in the United States as well, by letting on that our UFOs behaved the same way as those in Britain.

It seems inconceivable that just a few years later, in 1997, U.S. officials *somewhere* would not have taken serious note of the similar UFO sightings in Arizona. Obviously, officials at the DoD responding to the 2000 court-ordered search were not the same ones who had made the inquiry to the MoD about the Cosford triangle. Likely, they would not have known anything about this earlier exchange. However, the British inquiry about the Cosford UFO was sent to "all interested agencies," which must have included *some* department within the DoD. Unfortunately, we do not have any way of determining the thoroughness of the DoD search, nor do we know from where the intriguing question about our own mysterious triangles, posed to the UK, originated within our government.

In 2000, during the court litigation, did the DoD make inquiries to authorities in other government departments as part of an effort to do everything possible to obtain information about these objects? Wouldn't it make sense that the DoD might even approach the UK in such a circumstance, as it had done before, to find out if that country had had similar occurrences? This is not what they were asked to do by the court, and we have to assume that this level of search and widespread communication did not occur. Still, it's hard to fathom how the DoD staff required to find out about the 1997 objects could have come up with absolutely nothing to address the concerns raised by the citizens of Arizona, and not have been concerned about public reaction to the incident.

If indeed the DoD did *not* have any information about the 1997 unidentified objects of unknown origin operating over the United States, anywhere within the department, this in itself is a remarkable state of affairs. Were officials there alarmed by the information provided by witness affidavits through the court, and did they want to find out more? Some might consider such disregard of a massive, intruding object hovering over an American state to be grossly irresponsible, especially by those in charge of defending our country. Others might consider it so inexplicable that they would speculate whether DoD personnel were instructed by emissaries from the "controllers" of UFO information within a secret black program to keep quiet. Perhaps things have changed since 9/11, for it now seems hard to imagine that such an apparently advanced technological object, stealthily evading radar detection, could travel silently over a capital city and escape notice by federal authorities. Nonetheless, to this day, U.S. officials continue to keep the lid on the Phoenix Lights and other American sightings of mysterious giant triangles that have occurred since.

The case simmered for the next seven years until former Arizona governor Fife Symington brought it into the limelight in 2007, at the time of the tenth anniversary of the Phoenix Lights. He unexpectedly made a dramatic surprise announcement: that he, himself—despite his spoof press conference while governor—had actually witnessed what he called a "craft of unknown origin" along with his fellow citizens on that same March evening, but had decided not to make this public. In addition, he stated that the case remained unsolved, that it should be officially investigated, and that UFO incidents in general need to be taken seriously by the U.S. government.

On that unforgettable March evening in 1997, Symington had already arrived home and was watching the news when he received some calls about the sighting. He jumped into his car, and without his usual security detail, which had just left, he drove to a park near Squaw Peak, outside Phoenix, and amazingly enough, saw something highly unusual, brightly lit, overhead. "It was dramatic," he said in our first interview. "And it couldn't have been flares because it was too symmetrical. It had a geometric outline, a constant shape."

A Harvard graduate and decorated Air Force veteran of Vietnam, Symington is a great-grandson of Henry Clay Frick, the coal and steel magnate, and a cousin of the late Stuart Symington, Democratic Senator from Missouri. He served as the Republican governor of Arizona beginning in 1991, and was reelected in 1994. A longtime pilot, he frequently flies his twin-engine Beechcraft Baron plane between his two homes in Phoenix and Santa Barbara, California.

Symington was first nudged into coming forward in late 2006, when my colleague James Fox, an accomplished documentary filmmaker, sent him a copy of his UFO documentary *Out of the Blue*, which includes coverage of the Phoenix Lights. Fox was adding new material to the acclaimed film for a second release. He had never spoken to the former governor and decided to approach him to see if he could find out why Symington had staged the infamous spoof press conference. Fox had interviewed numerous witnesses who did *not* think Symington's spoof was funny, and were still rather upset by what, to them, was the governor's mockery and ridicule. He assumed that, given this behavior, the conservative governor did not take UFOs seriously, and he had no expectation that Symington would agree to an interview.

When he received *Out of the Blue,* Symington watched it and apparently found it fascinating, but at first was hesitant to reply. Eventually he came around. At that point, Symington says, he decided that when he met with the filmmaker, he would tell him the whole story. "I was sick and tired of people being ridiculed for reporting legitimate sightings," he later explained to me, and he decided that it was time to take a stand. Still, James Fox had no idea what was in store when he first met the former governor in Santa Barbara, and started his cameras rolling.

The two men seemed to hit it off right away. At one point during the filmed interview, Fox pulled out his cassette tape recorder. While the camera held a close-up of Symington's face, capturing his subtle change in expressions, Fox played for Symington a personal message he had recorded from one of the governor's former constituents, Stacey Roads. Roads and her teenage daughter were witnesses to the Arizona UFO, and she began by describing exactly where they were when she saw the craft. "A massive triangle came over I-10 and over my car. It was so large that if I'd opened a newspaper and laid on my back I couldn't have blocked out the entire object. It was traveling very slow without any noise," she said on the tape. The recording continued as Fox asked Roads whether she had a question she would like to ask the governor, and she replied: "Is this still a matter of ridicule to him, after he came out on TV with his alien, making us all look a little foolish? We've all been unwavering on our descriptions and a lot of evidence has come out since then. Does he still feel this is a matter of ridicule or has he taken a new stance?"

Governor Symington responded immediately and thoughtfully, without the least bit of fanfare. "I never felt the overall situation was a matter of ridicule, although we certainly took advantage of it, no question about it," he admitted. "But I don't consider it a matter of ridicule. It was a legitimate occurrence; a craft of unknown origin; who knows where from; inexplicable, and probably one of the major sightings in modern history in the country, because so many people saw it in Maricopa County—and *I saw it, too.*"

James Fox was absolutely unprepared for such a response. "I was shocked," he recalls. "It took me a moment to process it. I was thinking, did I really hear what I think I just heard? My immediate impulse was to make sure the cameras had been running, and they were. I didn't want to press the point right away, but wanted him to feel at ease. I left and reviewed the tape. It took a day or two for this to really sink in, and for me to realize I had something huge here."

Having kept the Symington story under wraps for about six months, James Fox called me in early 2007 to tell me about it, because we were approaching the tenth anniversary of the Phoenix Lights, with commemorative events planned in Arizona. We discussed the possibility of breaking the story in the print media at that time, just in advance of the updated film's release, which included the original interview. Symington seemed pleased with the idea of having the first written piece about his witnessing the UFO presented by someone who understood the bigger issue and proper context for the story, and who would treat it with respect. As a journalist, I was of course delighted with this "scoop" and knew that the mainstream media reporters would run with it afterward, including those who had made light of the incident in the Phoenix press years ago. But this time, they would be forced to read a proper, well-researched, serious piece before they could grab the news for themselves. This was an opportunity, although fleeting, for me to present a breaking UFO story in the way it should be told.

I was introduced to Symington via telephone, and conducted a long interview in which he expanded on what he had said to James Fox. I was struck by his sincerity, and although he was now a relatively private man who had no further interest in running for political office and did not relish exposure in the media, he voiced his commitment to helping both James and me in our efforts to bring greater credibility to the subject of UFOs and to impact government policy.

On March 18, 2007, I broke the Symington story in a front-page article for a relatively small Arizona newspaper, *The Daily Courier*, headlined "Symington Confirms He Saw UFO 10 Years Ago." I selected the *Courier* because it had a past record of providing good, fair coverage of the Phoenix Lights. As anticipated, the story had a dramatic impact and swept through national television newsrooms for days afterward, putting Symington in great demand. He made appearances on CNN and FOX News, but turned down all other requests.

Over the years, I've interviewed Symington several more times and come to know him. His remarkable personal journey as both a governor and a UFO witness, forced to contend simultaneously with the impact of his own sighting and the restrictive force of the UFO taboo on elected officials, is highly unusual. It certainly gives him a unique perspective, and has led him to become an advocate for change to an outdated and counterproductive UFO policy—or nonpolicy—in Washington.

But what makes Symington's situation even more exceptional is that although he was awestruck by his sighting, and believed this craft could not have been man-made, he didn't just simply ignore it. He went so far in the other direction as to stage a farcical press conference featuring a costumed alien that inadvertently insulted his fellow witnesses. How could he have laughed about this, and made a public joke out of it, given his direct experience of the physically real, inexplicable event a few months earlier?

Symington, in retrospect, says, "If I had to do it all over again I probably would have handled it differently." But the state of Arizona was "on the brink of hysteria" about the UFO flyover when he called the press conference, and the frenzy was building. "I wanted them to lighten up and calm down, so I introduced a little levity. But I never felt that the overall situation was a matter of ridicule," he says. That was why, ten years later, free of the constraints of political office, he wanted to set the record straight and make amends to constituents like Stacey Roads.

Now, we can gain insight from the former governor into what drives government officials to intensely resist the simple acknowledgment of the mere existence of something unidentified in the sky that does not *have* to be associated with anything extraterrestrial or alien. In this unusual case, the official knew it was real because he had seen it with his own eyes and didn't have to rely only on other witness reports. But hundreds of others also saw it! He still held back. How could he have restrained himself? And why did he feel compelled to do so?

He explains it this way:

You're not a normal person when you're a governor. You have to be extremely careful about public statements and how you handle yourself. A public figure is fair game for attack. Everything is picked over by the media and your political opposition. You try to avoid being the subject of harsh ridicule because you have a serious responsibility while in this role, and your public stature is directly related to your ability to get things done. If all of a sudden you're typed as a buffoon or a loony, you won't be effective. I had to make a choice. My top priority was to fulfill the responsibilities I had been elected to accomplish as governor.

In the months following the event, Symington had observed the press making fun of his friend Frances Barwood for simply taking the sighting seriously in response to public pressure—and she wasn't even a witness. He was also dealing with his share of political battles within the vicious world of Arizona politics, and says today, "Can you imagine what would have happened if I had said anything?" Although his decision is understandable, this is a sad commentary on our unspoken political policy toward UFOs, and the power of that irrational, habitual taboo that most of us have not questioned and that led Governor Symington to believe he would be branded a "buffoon" or a "loony" if he acknowledged something he and countless others had seen in the sky. Although he was at risk politically, such damaging labels are not only dangerous for political figures such as he, but are also harmfully applied to many everyday people who witness the phenomenon. Imbued with prejudice and an irrational fear of the unknown, these attitudes have been entrenched in our culture for over fifty years, and have not been well understood. But Symington's experience, for one, shows why elected officials and military brass in America wait until after retirement before risking saying anything at all about UFOs, no matter what their experience.

At the time, this governor found himself facing an unprecedented situation. Suddenly confronted with an escalation of public outcry following unanticipated national news coverage of a state-wide UFO sighting, he had to act fast. He felt it had become urgent that he change the mood. His administration was on its own in that moment, with no idea what had passed through the skies over Phoenix, or how to handle the aftermath of this momentous event. There was no support coming to state officials from the federal government, no answers from local authorities, and public ridicule had been unleashed against those daring to question what happened. So, relying on his own personal strengths in dealing quickly with a highly unusual problem, Governor Symington opted for a public spoof to lighten things up and cut the momentum with one sharp blow. "I never felt this sighting represented any kind of a threat," he explains. "I also had a good sense of humor. Everyone, including the media, was caught off guard. This seemed like an effective way to change things."

Imagine, for a moment, if a government office tasked with the investigation of UFO events such as this—exactly what we're hoping to establish now—was in place at the time of the Phoenix Lights, and the case had been properly handled. One can envision the following: During the actual event, as the result of a few calls from Washington, pilots already aloft could have been asked to fly near the objects, observe them, and photograph them if possible. Air Force jets would have been scrambled to get a closer look and attempt to engage the objects further. Civilian and military air traffic controllers could have attempted to catch them on radar, and military bases could have tried to contact the objects via communications signals sent out from the best technology for doing so. High-powered telescopes would have been aimed at the skies, at the proper altitude to possibly view the objects. The lead investigator from our UFO desk would have been in phone contact with a local team of scientists and aviation experts, already on the ground in Arizona or nearby states as part of an established network.

Early the next morning, the official from our agency would be dispatched to Phoenix for a briefing with all relevant officials, including, of course, the governor. His own sighting, and perhaps those of other officials or their families, along with commercial and military pilots, would be discussed and documented. Civilian witnesses would be encouraged to file independent reports and supply drawings of what they saw, along with any photos or home videos, as quickly as possible. Reporters would supply footage and witness interviews captured on camera the previous evening. Our coordinating official from the central office would have access to all radar records, and could interview air traffic controllers, police officers, government offices receiving calls, and all pilots flying near the multiple objects. Air Force bases and military installations in Arizona—all having been put on alert during the flyover—would be approached regarding the object, and would inform the investigators whether any flare drops, unusual flight formations, or other military maneuvers had been scheduled that night.

The public would be informed through a series of press conferences—like those provided by the National Transportation Safety Board (NTSB), as an example, in the days following an airline crash—about progress in the investigation. Citizens would be assured that the sighting did not constitute a threat, that no one had been harmed, that the proper authorities were investigating the incident, and that the public would be kept abreast of developments. Ideally, this event would not be sensationalized or blown out of proportion by the media, and would simply be one of many news stories of the day, perhaps not even of interest to the many who didn't witness an unidentified object themselves.

In short, a small agency, with links to experts within multiple disciplines around the country, could undertake a clean, clear, and thorough investigation of something like the Phoenix Lights within a short time frame. If the identity of the objects could not be determined after a reasonable amount of time, there would be no need to withhold that from the public. People would go about their lives, as they have done in Europe and South America when such announcements were made, and the scientific community—by now actively investigating the phenomenon—would be provided the relevant data for further study.

"If the sighting affecting so many people in Arizona could have been officially, quickly, and openly investigated, with no stigma attached, all the resulting public confusion and hysteria that I faced as governor could have been avoided," Symington states. "This is the sane approach, as is recognized in other countries, and should become the new American policy. I would not want to see another governor go through what I did in 1997, and it's only a matter of time before this will happen again."

No wonder apprehension and frustration mounted in the state of Arizona. How could anyone feel safe, or trust the authorities to protect them, when such an intrusion by a massive craft is treated as if it never happened? Each of us must ask ourselves what *we* would have done, and how *we* would have felt, if we had stood beneath this silent hovering object. It makes enormous sense to have a small agency in place to be prepared for the eventuality of another widely witnessed UFO event.

Another factor, as has been pointed out by many military officials, is the risk that potentially disastrous aggressive actions might be taken against a UFO, due to a lack of preparation of those responsible for the defense of the country. If an object the size of the one seen over Phoenix came even closer to the ground, for example, or shot a penetrating beam onto an observer, or took any number of frightening actions that we could imagine, how would we respond? Pilots have attempted to shoot down UFOs from the air. What would it take for a similar response to be triggered from an air defense base on the ground? We must not forget that we are dealing with something so unknown to us, so entirely unexplained, that we have no idea what could happen the next time one appeared. The establishment of a government office would be the first step in the distribution of the appropriate data, preparation manuals, and policy recommendations to the Air Force and all other military installations around the country.

The state of Arizona has seen more than one prominent elected official confront the UFO problem. Prior to his sighting, Fife Symington had enjoyed a long-term relationship with a mentor who had strong opinions about U.S. government secrecy and UFOs. Barry Goldwater, five-term senator from Arizona, Republican presidential nominee in 1964, pilot, and friend of the Symington family, was a hero and a father figure to him beginning at age twelve. Goldwater served as campaign chairman for both of Symington's successful runs for governor.

Symington recounts that on a number of occasions, when he was flying to campaign events with Goldwater, the former senator told him about his efforts to obtain secret UFO information from Wright-Patterson Air Force Base, as Goldwater has written in his letters. It's interesting that Symington never knew that Goldwater had written anything about his ventures until after he recounted these conversations to me, when, much to his amazement and delight, I sent him copies of the letters. "Barry was convinced that UFOs exist and that the government held top secret stuff and was holding it close for technological reasons. He didn't *know* this as a fact, but he was highly suspicious," Symington says. Unfortunately, Goldwater was not well enough to comment on the Phoenix Lights incident, having suffered a stroke in 1996. He died in 1998 at his home outside Phoenix.

Today, Symington is inclined to agree with Barry Goldwater that our government is withholding secret information about UFOs. "If we got our hands on a very advanced spacecraft before anyone else, you can be sure we would hold it tight and work on it, and we would be interested in the advanced technology. This is as valid as any other idea to explain why it would be kept secret," he says.

Governor Symington's "coming out of the closet" represents a historical turning point in the effort to bring official recognition and policy change to the UFO issue in America. Never before has a twice-elected official of this stature not only acknowledged witnessing an unmistakable unidentified flying object, but also taken a public stand advocating for change. When he was forced to test the system, the governor discovered firsthand that it doesn't work. As a result, he has to some extent made this effort a personal mission, which is being carried forward with the support of other equally convinced former officials from other countries, some of whom have come together in this volume. As a former elected government official in America and part of the political establishment, Symington is uniquely positioned to influence a change in policy. Through his contacts and experience in government, he can help move us toward the founding of a new government agency—which he could have benefited from so much while in office—and has already done so by adding his voice and support to our international coalition.

CHAPTER 25

Setting the Record Straight
by Fife Symington III
Governor of Arizona, 1991–97

Between 8:00 and 8:30 on the evening of March 13, 1997, during my second term as governor of Arizona, I witnessed something that defied logic and challenged my reality: a massive, delta-shaped craft silently navigating over the Squaw Peak in the Phoenix Mountain preserve. A solid structure rather than an apparition, it was dramatically large, with a distinctive leading edge embedded with lights as it traveled the Arizona skies. I still don't know what it was. As a pilot and a former Air Force officer, I can say with certainty that this craft did not resemble any man-made object I had ever seen.

As soon as I reached home I told my wife, Ann, about it. She listened attentively, and we both thought long and hard about whether I should make public what I had seen. Eventually, at least for the time being, we reached the conclusion I should not, as doing so would most likely result in ridicule from the press that would distract me and my entire administration from the work we had been elected to accomplish.

The same incident was witnessed by hundreds if not thousands of people in Arizona, and my office was at once besieged with phone calls from concerned Arizonians. Even so, I managed to keep my head down—until two months later, when a story about the sightings appeared in *USA Today*. Catalyzed by the article, hysteria intensified to a point that I decided to lighten the mood and add a note of levity by calling a press conference at which my chief of staff arrived in alien costume. Originally my idea, this was one my team immediately embraced with enthusiasm. Not only would it dampen any incipient panic, it would show the human face of those who hold public office.

In the event, we did manage to calm the public's developing anxiety and, despite the fact that, in the process, we also upset a few of my constituents, I felt that our approach had ultimately served a greater good.

With hindsight, however, I would like to set one part of the record straight. As I assured James Fox when he interviewed me for his documentary film, *Out of the Blue*, it was never my intention to ridicule anyone. My office *did* make inquiries—of the Department of Public Safety, the Air National Guard, and the lead officers at Luke Air Force Base—as to the origin of the craft, but to this day all of these remain unanswered.

Eventually, the Air National Guard claimed responsibility, stating that at the time its pilots had been dropping flares. This explanation, however, defies common sense, for flares do not fly in formation. Indeed, such a narrative seems indicative of the attitude one all too often encounters in official channels, which provide ex post facto rationales—e.g., weather balloons, swamp gas, and military flares—apparently meant to accord with our experience and expectations rather than our observations.

I was never satisfied by this silly explanation. For, although, as suggested by analysis (by Dr. Bruce Maccabee, among others) of a video taken then, there may well have been military flares in the sky later that evening—around ten o'clock, to be exact—what I and so many others observed between eight and eight-thirty was, on inspection, something else entirely: a huge and mysterious craft.

Today, of course, I know that I was not alone in having witnessed something so extraordinary. There are many high-ranking military aviation and government officials who have witnessed similar apparently inexplicable things at other times and in other quarters of the sky, and who share my concern that our government disparages these facts at its, and our, peril. Some of them have come together in this book, and I join them in suggesting a new approach to this problem.

With due respect, we want the United States government to cease perpetuating the myth that *all* UFOs can be explained away in down-to-earth, conventional terms. Instead, our country needs to reopen the official investigation it shut down in 1970. We should no longer shun international dialogue on this important subject. Rather, we urge the appropriate agencies of our government to work in cooperation with countries that have already begun exchanging reports of sightings and to endeavor, in a spirit of genuinely open scientific inquiry, to learn more about UFOs and to make the results of such inquiries, whether immediately comprehensible or not, fully public.

CHAPTER 26

Engaging the U.S. Government

In 2002, I cofounded the Coalition for Freedom of Information, an independent alliance and advocacy group whose mission is to achieve scientific, congressional, and media credibility for the often misunderstood subject of UFOs. Much of our work has been built around an effort to acquire new information through the Freedom of Information Act, and it quickly won the support of John Podesta, one of our country's strongest advocates for openness in government, who contributed the foreword to this book. As President Clinton's former chief of staff, Podesta was instrumental in the declassification of 800 million pages of documents during the Clinton administration. In 2008, he headed President Obama's transition team and now directs the preeminent Center for American Progress in Washington. Our FOIA initiative resulted in the settlement of a federal lawsuit against NASA in our favor, requiring the agency to release hundreds of pages of previously withheld documents.

The coalition is asking for responsible action on the part of the United States concerning UFOs. We make this request *not* as an accusation of wrongdoing in the past, but as an invitation to join an international, cooperative venture under way now. In petitioning for such a change, as previously described in relation to the Phoenix Lights incident, we are seeking the creation of a small government agency to investigate UFO incidents, and to act as a focal point for action at home and for research worldwide. Through its legitimization of the subject, such an agency would stimulate scientific interest and assist with the allocation of government and foundation grants for interested scientists in the academic, research, and aviation communities. As the work of the agency develops over time, positive attitudes toward the serious study of UFOs would be nurtured, leading to the liberation of additional resources. Public support—already very strong although without a focal point—would grow for a global research project that could ultimately solve the UFO mystery.

The first step in approaching a member of Congress or the Obama administration to facilitate this endeavor is to make it clear, as we have continuously done in these pages, that a UFO is, by definition, simply something unidentified. The agnostic position, the scientifically sound one, acknowledges the accumulated evidence of some kind of physical, extraordinary phenomenon but recognizes that we do not yet know what it is. The proper understanding of the acronym "UFO" must lie at the heart of any approach to the American government if it is to be successful, and the necessity of that simple adjustment in understanding—ending the automatic equating of UFO with extraterrestrial spacecraft—cannot be overestimated. This would lay a foundation that would allow politicians to be able to publicly consider moving forward with this issue. This may be obvious to most readers, but some activists working for change do not make this important distinction. Instead, they make sometimes outlandish claims about UFOs and related government conspiracies that cannot be substantiated—and they still expect to be taken seriously. No matter what anyone's personal beliefs are about the nature of UFOs, those in high positions—the only ones capable of effecting real change—are obviously not going to accept any explanation before a new, legitimate scientific investigation makes a definitive determination.

The need for a new way of thinking about UFOs was painfully illustrated when NBC's Tim Russert popped a surprise question to Ohio congressman Dennis Kucinich during the nationally televised presidential debate in 2007. Russert asked Kucinich whether he had actually seen a UFO, as was reported in a book by Shirley MacLaine. Snickers from the studio audience became audible as soon as the dreaded U-word was uttered. The poor man replied, accurately, that yes, he had simply seen something unidentified, reiterating that it was "an unidentified flying object." Despite the straightforward honesty and clarity of his reply, Kucinich could not escape the laughter that had begun even before he had a chance to speak. He followed his comment with a joke of his own, as a way of saving face.

A U.S. government office, like the British UFO desk or the French GEIPAN, would quickly dispense with the notion that this subject is silly. We need a different language, a whole new frame of reference without the baggage of the past. Some scientists and military officials have attempted to begin this process by switching to the broadly defined term "unidentified aerial phenomena," or "UAP." This obviously is not enough to change the deeply embedded association of UFOs with science fiction or mental aberrations, but for them it is a step in that direction, and also helps to lessen the power of the taboo.

A small, simple change in policy is all it would take to make a very big difference. A body within the government to address the UFO issue can be set up easily, quietly, and inexpensively. To get started, all it requires is funding for a small office, staffed by one to three people, equipped with a few computers and file cabinets and tucked away in one of many possible locations. The staff would create links to scientists, law enforcement officials, civilian researchers, and specialists from a range of disciplines, who would step in as needed if a major UFO event were to occur. Few additional resources would be necessary, because investigation of the occasional worthwhile cases would involve drawing on established facilities, equipment, and personnel, such as cross-referring to satellite imagery and existing records of aviation, meteorological, astronomical, and radar data. Reputable labs could be used for the analysis of photographic images and physical evidence. A qualified volunteer board of advisors, to include academics, scientists, and retired military officers, would meet regularly with the staff to offer input and help coordinate the public release of information. Ideally, information about UFOs that may currently be withheld by U.S. intelligence agencies would be released to the office and the public.

Details of the mission and structure of the agency would obviously have to be carefully worked out, but experienced people are ready and available to assist in that process and make sure the mistakes of Project Blue Book are not repeated. This new plan would initiate a fundamentally different organization from that of Blue Book, because it would be committed, with public oversight, to properly investigate cases and to work with other countries. It would be the opposite of our previous Air Force agency—a controlled public-relations mechanism covering up the unsolved cases—that existed in the 1950s and '60s.

In November 2007, twenty-two distinguished individuals, including six retired generals, from eleven countries signed a formal request for such an agency to be established. The "International Declaration to the United States Government," which I drafted in cooperation with members of my group, the Coalition for Freedom of Information (CFi), includes most of the writers for this book along with five others, and is posted on the CFi website. The document is signed by current and former military and government officials and pilots, each of whom, while on active duty, "has either been a witness to an incident involving an unidentified flying object or has conducted an official investigation into UFO cases relevant to aviation safety, national security, or for the benefit of science."

The declaration states that the current level of disengagement by the American government with important UFO sightings, such as the Phoenix Lights and the O'Hare sighting, "represents both a missed opportunity and a potential risk." The call to action asks the U.S. government to "join in cooperation with those governments which, recognizing the reality of unidentified flying objects and related aviation safety concerns, have already set up their own investigative agencies." It suggests that the U.S. Air Force or NASA serve as the location for such a research effort and ends with a final request: "We call on the United States of America to engage with us and with currently active officials around the world to address this problem in an ongoing dialogue."

The credentials of the names making this request are impressive. As a result, the document received wide coverage in the press when it was endorsed by former governor Fife Symington and released at the November 2007 press conference in Washington, D.C. But nothing has changed as a result. Our group sidelined this initiative during the build-up to the 2008 presidential election, which fully occupied the country, and in the time following when the new Obama administration first took office and was faced with numerous engrossing and urgent challenges. Yet we remain as convinced as ever that this is not too much to ask. It is something the American public has wanted for a long time, and now that we have an administration committed to openness and a global vision, with a commander in chief who is also a Nobel Peace Laureate, our chances of success are better than ever.

CHAPTER 27

Militant Agnosticism and the UFO Taboo
by Dr. Alexander Wendt and Dr. Raymond Duvall

In August 2008, I received an e-mail from Dr. Alexander Wendt, a professor of political science at Ohio State University; he attached his twenty-six-page paper just published in the leading scholarly journal Political Theory. *Co-authored with Dr. Raymond Duvall, "Sovereignty and the UFO" provided a complex, detailed, and deeply thoughtful analysis of why governments systematically ignore the UFO phenomenon despite the overwhelming evidence for its existence. We've touched on various aspects of the UFO taboo within these pages, exploring also the question of secrecy and possible threatening aspects of UFO reality, but even so, the deeper questions remain unanswered: Despite all the evidence, why is the prohibition against taking UFOs seriously so powerful, and what keeps it going? In order for a new government agency to function properly and successfully, this is the final aspect that must be addressed along with the logistical and structural proposals.*

In my many years of work with this material, the unresolved loose ends involving issues related to the UFO taboo seemed to point to something larger and more fundamental than had been articulated, but it wasn't clear what that was. Former Air Force scientific consultant J. Allen Hynek probed this question in 1985, but was unable to resolve it. He described the problem as a strange "malady" with the power to plunge its victims into "a deadly stupor. Like a virulent apathy virus, it could easily immobilize cities and the entire country . . . as though a bad fairy had administered a sleeping potion." Yet he couldn't quite find the reason why it so severely afflicted those responsible for running governments and protecting citizens, and therefore he could not offer a cure.

Now, the same question has been taken up by two accomplished political scientists, putting fresh eyes on the problem from within the aca-

demic community. Alexander Wendt is the author of the award-winning book Social Theory of International Politics *(Cambridge University Press, 1999), and is interested in philosophical aspects of social science and international relations. Raymond Duvall is professor and chair of the Department of Political Science at the University of Minnesota. His focus is on critical theories, with particular attention to power, rule, and resistance in world politics. The two met when Alexander Wendt was a student of Duvall's while in graduate school, and they have remained in touch since then. Beginning around 1999, Wendt spent about five years reading and thinking about the UFO subject on his own. "I tried to figure out what's really real in this context, given how much non-sense, disinformation, and conspiracy theorizing there is out there," he told me.*

In 2004 he started talking to his former advisor about his ideas and their relevance to political theory, and the decision to explore the taboo emerged from these discussions. "I initially approached him with a focus on why there was official secrecy about UFOs," Wendt explains. "Talking with him helped me see that secrecy was just a symptom of the problem, which goes much deeper." At first, Duvall was skeptical at best, he says, having given no thought to UFOs before Wendt initiated a conversation about them. "It's probably fair to say that I embodied the taboo," he wrote in an e-mail. "Working on this paper with Alex has transformed my thinking."

The two scholars deconstruct the arguments made by debunkers that perpetuate the cultural and political position that UFOs should not be taken seriously, and they examine the deep-seated fear of the extraterrestrial hypothesis that underlies such irrational skepticism. Yet, ironically, they say that they were directly impacted by this very taboo themselves after publishing "Sovereignty and the UFO." In this sense, the paper became a "natural experiment," providing a textbook illustration of their thesis. "As the first article taking UFOs seriously published in a social scientific journal in decades—if ever—one might have expected it to generate some controversy," Wendt says. "Academics certainly get into controversies about much less, and they are usually interested in debating such papers. But to our knowledge, none of our fellow social scientists, in the English-speaking world at least, has yet taken up the paper's challenge. This is disappointing, but this dismissal is at least consistent with the paper's hypothesis that there is indeed a taboo on this topic which prevents reasoned debate."

Dr. Wendt and Dr. Duvall agreed to write a new essay specifically for this volume, incorporating their ideas from the first article into one designed for nonacademic readers, with some new thoughts added. I hope this piece will help address lingering questions about the roots of the fundamental disconnect between the powerful evidence for UFOs and the disinterest of our government and scientists toward investigating them. It should also disarm the debunkers who routinely come up with defensive arguments that show they have not actually studied the facts, in itself an illustration of the taboo in action. Since the paper distills these arguments and dispenses with them, perhaps we can all gain a new perspective on these debunkers and adopt a more rational approach to the disconcerting questions raised by the mystery of UFOs.

There is a taboo on this book—the UFO taboo. Not in popular culture, of course, where interest in UFOs abounds and websites proliferate, but in elite culture—the structure of authoritative belief and practice that determines what "reality" officially is. With respect to UFO phenomena this structure is dominated globally by three groups: governments, the scientific community, and the mainstream media. Although their individual members may have varying private beliefs about UFOs, in public these groups share the official view that UFOs are not "real" and should not be taken seriously—or at least no more seriously than any other curious cultural belief. For these elites, a book like this, which *does* take UFOs seriously, is intrinsically problematic.

One manifestation of the UFO taboo is official disinterest in responding to UFOs or in finding out what they are. Since 1947, when the modern UFO era began, neither the scientific community nor governments (with the partial exception of France) have made a serious effort to determine their nature, as far as we know. Reports have been filed and a few officially investigated after the fact, but the vast majority have been ignored, and no authoritative effort has been made to survey systematically or seek out UFO phenomena. The media reinforce this disinterest by rarely covering UFOs, and when they do it is inevitably with a wink and a nod, as if to reassure us that they don't *really* take UFOs seriously, either.

Given that modern science seems to find almost everything in nature interesting, such disinterest is puzzling. But disinterest alone does not make a taboo—which is something prohibited, not just ignored. Rather, what gives the UFO this special status is that it is considered to be outside the boundaries of rational discourse. Although members of the general public might *believe* that UFOs exist, the authorities "*know*" that UFOs are merely figments of overactive imaginations, no more real than witches or unicorns. Thus, to take UFOs seriously is to call one's *own* seriousness into question. When UFO "believers" appear to deny empirical reality, there is not much more for the elite culture to do than either ignore or condemn them as irrational or even dangerous. In this light the UFO appears not as an "object" at all, but as a troublesome fiction that is best not talked about—in short, a taboo that prevents reasoned debate.

Yet, the reality is that UFOs are *not* matters of belief, but facts. Many thousands of reports worldwide describe unexplained objects in the sky. Most consist only of eyewitness testimony, which might be disregarded as unreliable—and some undoubtedly actually are—but the fact that many UFO reports come from "expert witnesses" like commercial and air force pilots, air traffic controllers, cosmonauts, and scientists should give one pause. However, some UFO reports are also corroborated by physical evidence, including scientifically analyzed photo and video images, physical ground traces affecting plants and soil, effects on aircraft, and anomalous radar tracks. In modern society, physical evidence is normally considered definitive evidence of reality, objective evidence for *something* that has a cause in the physical world. By this criterion, then, at least some UFOs are clearly real. The question that makes them a problem is: Could they be extraterrestrial?

Proving Our Ignorance

UFO skeptics think that human beings know, as a matter of scientific fact, that UFOs are *not* extraterrestrial and therefore can be ignored. Yet none of the strongest arguments for this view in fact justify rejecting the extraterrestrial hypothesis as a possible explanation for UFOs. They don't even come close. Actually it is *not* known, as a matter of scientific fact, that no UFOs have an extraterrestrial origin. If we reject this hypothesis anyway, we are rejecting what just *might* be the true explanation, without having submitted it to the test. Again, this does not mean that UFOs *are* extraterrestrial, either; UFOs are, after all, *un*identified. But that is precisely our point: At this stage human beings simply do not know.

Given that little systematic science has been done, the case for rejecting the extraterrestrial hypothesis out of hand rests on an a priori theoretical conviction that extraterrestrial visitation is impossible: "It can't be true, therefore it isn't." Skeptics offer four main arguments to this effect.

"We Are Alone." Human beings have debated for centuries whether intelligent life exists elsewhere in the universe, and with the recent discovery of over 400 extrasolar planets, this debate has heated up considerably of late. Good scientific reasons exist to think that intelligent life does not exist elsewhere, but increasingly there are equally good scientific reasons to think that it does. Bottom line: We don't know yet.

"They Can't Get Here." Skeptics argue that even if there is intelligent life elsewhere, it's too far away from Earth to get here. Relativity theory tells us that nothing can travel faster than the speed of light (186,000 miles per second). At .001 percent of light speed, or 66,960 miles per hour—already far beyond current human capabilities—it would take 4,500 Earth years for any vehicle to arrive just from the nearest star system. And at speeds much closer to light a single spaceship would need to carry more energy than is presently consumed in an entire year on Earth.

Physical constraints on interstellar travel are often seen as the strongest reason to reject the extraterrestrial hypothesis, but are they clearly decisive? Computer simulations suggest that even at speeds well below light, any expanding advanced civilizations should have reached Earth long ago. How long ago depends on what assumptions are made, but even pessimistic ones yield encounters with Earth within 100 million years, barely a blip in cosmic terms. Additionally, there are growing doubts that the speed of light is truly an absolute barrier. Wormholes—themselves predicted by relativity theory—are tunnels through space-time that would shorten greatly the distances between stars. And then there is the possibility of "warp drive," or engineering the vacuum around a space-ship to enable it to skip over space without time dilation. Such ideas are highly speculative, but given how far we humans have come in just 300 years since our scientific revolution, imagine how far another civilization might have advanced 3,000 years (much less 3,000,000) after theirs. In light of these arguments, if anything, visitors from other civilizations *should* be here, which prompts the famous "Fermi Paradox," or "Where are they?"

"They Would Land on the White House Lawn." So skeptics often take the argument one step further, by asking: If visitors from other planets have come all this way to see us, why don't they land on the White House lawn and introduce themselves? After all, if human beings were to encounter intelligent life in our own space exploration, that's what *we* would do. On this basis, the fact that UFO occupants have not done so is evidence that they are not here.

But is it? It is not at all clear that space-faring humans *would* land on an alien equivalent of the White House lawn if they journeyed to a distant planet. Perhaps advanced explorers would maintain a policy of non-interference toward lower life forms. Regardless of what human beings might do, however, on what *scientific* basis can we know the intentions of alien beings, whose nature and agendas might be utterly unimaginable to us? There is none, and as such one cannot rule out the possibility that extraterrestrials might have reasons for avoiding contact.

"We Would Know If They Were Here." This final argument appeals to human authority—that, due to our vast surveillance of the skies with sophisticated radar and telescopes, the world would know definitely by now if extraterrestrials were here, because the experts would have discovered them.

This position, too, is by no means decisive. First, it assumes an ability to observe and recognize UFOs that may be unwarranted; if some are vehicles able to visit Earth, then their occupants could easily have the technology to limit knowledge of their presence. Second, the authorities have not actually looked for UFOs, and what is not looked for or expected is often not seen. Finally, in view of pervasive official secrecy about UFOs, more is probably known about them than is publicly acknowledged. This does not mean that what is known is their origin, but in the face of so much secrecy it is natural to raise the question.

Importantly, our point about each of these arguments is not that they are wrong, but that *reasonable people can disagree* about whether they are wrong, since they all ultimately rely on unproven assumptions rather than established scientific facts. Indeed, the very fact that it is so easy to raise reasonable objections to UFO skepticism is further evidence that, scientifically speaking, human beings can't rule out the extraterrestrial hypothesis. Some of us may look at the evidence and arguments and conclude that the probability is zero, while others may give the hypothesis more credence—but who really *knows*? No one knows, because we do not have the scientific knowledge to make such probabilities meaningful. As former secretary of defense Donald Rumsfeld might put it, we are dealing here not with "known unknowns" but "unknown unknowns," where objective likelihoods are anyone's guess. And when there is such "reasonable doubt," scientific hypotheses should not be rejected a priori. Far from proving that UFOs are not extraterrestrial, in short, current science proves only its ignorance.

The Threat of the UFO

If the proper application of science demands that at present we be agnostic about whether any UFOs have an extraterrestrial origin, neither believing nor rejecting this, then the taboo on trying to find out what UFOs are is deeply puzzling. After all, if any UFOs *were* discovered to be from somewhere else in the universe, it would be one of the most important events in human history, making it rational to investigate even a remote possibility. It was just such reasoning that led the U.S. Congress for a time to fund the SETI program looking for evidence of life around distant stars. So why not fund the systematic study of UFOs, which are relatively close by and at least sometimes leave physical evidence? Even for those for whom the question of extraterrestrials is not on the table, what about simple scientific curiosity? Why *not* study UFOs, just like human beings study everything else?

Our thesis is that the origins of this taboo are political. As political scientists, we are concerned with a possible connection between the need to dismiss the UFO and the way in which modern peoples organize and govern their societies. The inability to see clearly and talk rationally about UFOs seems to be a symptom of authoritative anxiety, a socially subconscious fear of what the reality of the UFO might mean for modern government.

The threat is threefold. On the most obvious level, acceptance of the possibility that the UFO is *truly* unidentified, and that therefore an unknown, very powerful "other" might actually exist, represents a potential physical threat. Clearly, if some other civilization has the ability to visit Earth, then it has vastly superior technology to human beings, which raises the possibility of colonization or even extermination. As such, the UFO calls into question the state's ability to protect its citizens from such an invasion. Second, governments may also be reacting to the possibility that a confirmation of extraterrestrial presence would create tremendous pressure for a world government, which today's territorial states would be loath to form. The sovereign identity of modern states depends on their difference from one another. Anything that required subsuming this difference into a global sovereignty would threaten the fundamental structure of these states, quite apart from the risk of physical destruction.

Third, however, and in our view most important, the extraterrestrial possibility calls into question what we call the anthropocentric nature of modern sovereignty. By this we mean that, in the modern world, political organization everywhere is based on the assumption that only human beings have the ability and authority to govern and determine our collective fate. Nature might throw us a curve ball in the form of a pandemic or global warming, but when it comes to deciding how to deal with such crises, the choice is ours alone. Such anthropocentrism, or human-centeredness, is a modern assumption, one less common in prehistoric and ancient times, when Nature or the gods were considered more powerful than human beings and thought to rule.

Significantly, it is on this anthropocentric basis that modern states are able to command exceptional loyalty and resources from their subjects. Because a possible explanation for the UFO phenomenon is extraterrestrial, taking UFOs seriously calls this deeply held assumption into question. It raises the possibility of something analogous to the materialization of God, as in the Christians' "Second Coming." To whom would people give their loyalty in such a situation, and could states in their present form survive were such a question politically salient? Our contention is that the political survival of the modern state depends on that question *not* being politically salient. As such, an authoritative taboo on the UFO is functionally necessary for rule to be sustained in its present form.

In sum, the UFO creates a deep, unconscious insecurity in which certain possibilities are unthinkable because of their inherent danger. In this respect the UFO taboo is akin to denial in psychoanalysis: the sovereign represses the UFO out of fear of what it might reveal about itself. There is therefore nothing for the sovereign to do but turn away its gaze—to ignore, and hence be ignorant of the UFO—and make no decision at all.

Maintaining the Taboo

The suggestion that the UFO taboo is functionally necessary for modern, anthropocentric rule does not mean that it will be automatically maintained. Such a strong prohibition takes work. To be clear, this is not the conscious work of a vast conspiracy seeking to suppress "the truth" about UFOs, but the work of countless undirected practices that help us "know" that UFOs are not extraterrestrial and can therefore be disregarded. The work of the UFO taboo is paradoxical, however, because unlike the days when the visions of shamans and prophets were taken to be authoritative, in the modern world we know things by making them visible and trying to explain how they work—which in the UFO case would be self-subverting because it could lead to a validation of the extraterrestrial hypothesis. So what are needed are techniques for making UFOs "known" without *actually* trying to find out what they are. One might distinguish at least four ways of doing this.

The first is authoritative representations, or descriptions of what UFOs are, as provided by those having the authority to stipulate what defines official reality—governments, the scientific community, and the media. Four such current representations are especially noteworthy: (1) that UFOs are known by science to have conventional explanations, for all the reasons we criticized above; (2) that UFOs are not a national security concern, which allows states to wash their hands of the problem; (3) that any study of UFOs is by definition pseudoscience, since UFOs do not exist; and (4) that UFOs are science fiction, which displaces the existentially scary aspect of a potential extraterrestrial encounter into the safety of the imagination. We are not saying that modern authorities are consciously trying to protect the UFO taboo when they make such representations. Our point is that whatever the concrete intent in particular instances, these representations (and no doubt others) have the *effect* of reinforcing the authoritative consensus that UFOs should not be taken seriously.

A second technique by which the taboo is maintained turns the point about pseudoscience on its head. Here we are thinking of officially sanctioned but problematic inquiries into UFOs like the 1968 Condon report, the purpose of which was to give the appearance of an objective, scientific assessment while reaffirming the dominant view that there is nothing to such phenomena. As has been amply documented in the literature, in the Condon case this ideological bias led to gross errors of research design and empirical inference, as well as to an Executive Summary that completely rejected the extraterrestrial hypothesis even though conventional explanations could not be found for fully 30 percent of the cases that had been studied. This is not to say there is no good science in the Condon report (on the contrary), but that ultimately it was a "show trial" for the extraterrestrial hypothesis. Nevertheless, the report's conclusion that UFOs are definitely not extraterrestrial was immediately accepted by the larger scientific community, and also enabled the U.S. Air Force to disengage publicly from the UFO problem, which it had wanted to do for some time. That such a flawed report could be embraced so readily attests to how deep-seated the "will to disbelieve" is.

A third factor sustaining the taboo is pervasive official secrecy about UFO reports involving military personnel, the effect of which is to remove from the system knowledge that might bolster the argument for taking UFOs seriously, thereby (at least implicitly) reinforcing the skeptical case. UFO secrecy takes at least two forms. The most obvious is withholding information on known cases, whether by redacting text or telling citizens requesting documents through the Freedom of Information Act that no relevant documents exist at all. (In the United States, the law requires government agencies to inform the public if requested documents are classified, or else release them with sensitive sections redacted.) The other form of secrecy—not reporting military UFO encounters at all—is more difficult to assess, since it is impossible to know how many such cases there are. Still, the fact that most governments do not release UFO reports as a matter of course—although in recent years this trend has started to shift in some countries, but not in the United States—does not inspire confidence that we know the complete universe of cases.

This secretive pattern of behavior is of course grist for the mill of conspiracy theorizing, since it naturally raises the question "What is the government trying to hide?" However, we are concerned not with the particular content but only the *effect* of official secrecy, which helps to reinforce the UFO taboo by removing potentially contrary knowledge from the system. Our personal view is that far from hiding the truth about aliens the state is more likely hiding its ignorance, but who knows? In a context of UFO secrecy, personal belief is all we have.

The last mechanism is *discipline*, by which we mean techniques for ordering thought and action that rely not on rational appeals to science, but more nakedly on social pressures and power. A particularly prominent form in the UFO context is the social dismissal of people who express public "belief" in UFOs—through ridicule, gossip, shunning, public condemnation, and/or character assassination—so that it is not just the idea of UFOs that is dismissed but the person advocating the idea whose credibility is called into question. Given individuals' desires for approval, reputation, and professional advancement, an expectation of this kind of discipline leads to self-censorship, fueling the "spiral of silence" about UFOs that makes it so hard to speak out in the first place.

Resistance Through Militant Agnosticism

These are powerful mechanisms, and as such some might say that with respect to the UFO taboo, "resistance is futile." Yet the taboo has at least three weaknesses that make it, and the anthropocentric structure of rule that it sustains, potentially unstable.

One is the UFO itself. Despite authoritative efforts to deny their reality, UFOs stubbornly keep showing up, generating an ongoing need to transform them into non-objects. Modern governments might not recognize the UFO, but in the face of continuing anomalies, maintaining such nonrecognition requires work.

Another weakness lies in the different knowledge interests of science and the state. While the two are aligned today in authoritative anti-UFO discourse, ultimately the state is interested in maintaining its skeptical narrative about UFOs as certainly true, whereas science recognizes, at least in principle, that its truths can only be tentative. The presumption in science is that reality has the last word, which creates the possibility of scientific knowledge countering the state's dogma.

And then there is liberalism, the essential core of modern governance. Even as it produces rational subjects who know that "belief" in UFOs is absurd, liberalism justifies itself as a discourse that produces free-thinking subjects who might doubt it.

The kind of resistance that can best exploit these weaknesses might be called "militant agnosticism." By "agnostic" here we mean that no position on whether UFOs are extraterrestrial should be taken until they have been systematically studied. Resistance must be agnostic because, given our current knowledge, neither denial nor belief in the extraterrestrial hypothesis is justified; we simply do not know. Concretely, agnosticism means "seeing" the UFO for what it is rather than ignoring it, taking it seriously as a real and truly unidentified *object,* broadly defined to include any natural phenomenon. Since it is precisely such acknowledgment of UFOs' reality that the taboo forbids, "seeing" alone is a kind of personal resistance.

To be *politically* effective, however, resistance must also be militant, by which we mean public and strategic. Indeed, purely private agnosticism about UFOs, of the kind that people in the modern world might have about God, does nothing to break the spiral of silence that surrounds the issue and so in effect contributes to it. To break the cycle resistance needs to be directed at the central problem posed by UFO phenomena, namely reducing our collective ignorance about what they are, rather than at the side issue of official secrecy, which strategically is a diversion. (If we're correct that governments are hiding not the truth but their own ignorance, then even if they released all their files we would be no closer to knowing what UFOs are.) That is to say, what is needed above all else is a systematic *science* of UFOs, on the basis of which we might eventually be able to make informed judgments about them, as opposed to simply reiterating dogmas one way or the other.

To go beyond the minimal scientific research that has already been done and make new breakthroughs, such a science will have to do three things. First, it will need to focus on aggregate patterns rather than individual cases. Given our inability to manipulate or predict UFO phenomena, there are inherent limits to what case studies can show. Already, official analyses of selected cases have sometimes been able to rule out conventional explanations—what they are not—but this does not tell us what those UFOs *are*. UFOs are like meteorological phenomena, which can be properly studied only in the aggregate.

Second, a science of UFOs will need to focus on finding new reports rather than analyzing old ones. This is because existing high-quality reports are relatively few in number and were collected by accident and through a variety of means, making it almost impossible to find patterns. Moreover, there is only so much information that can be extracted from a historical report, particularly one disconnected from knowledge of the environmental context. Trying to generate new reports systematically might greatly increase our data points, and put them automatically into context, as well.

Finally, a science will need to focus on collecting objective, physical evidence rather than subjective, eyewitness accounts, for only the former will convince the authorities that UFOs "exist," much less that the extra-terrestrial hypothesis is worthy of consideration. Of course, getting such evidence is no easy task, but as shown by existing radar and video images, as well as chemical analyses of a few UFO "landing sites," it can be done.

Any serious attempt to satisfy these requirements will require considerable technological infrastructure (radar installations or other monitoring equipment) and large amounts of money. Normally one would expect the state to provide such capital. Although every effort should be made to bring this about, our particular theory of the UFO taboo—that it is a functional imperative of modern, anthropocentric rule—necessarily makes us pessimistic that world governments will act anytime soon. As such, it seems important strategically to consider, alongside efforts to enlist the state, alternative ways of establishing a science of UFOs.

Whether tackled by the state or by civil society, or both, the problem of UFO ignorance is fundamentally political before it is scientific, and as such a truly militant agnosticism will be necessary to overcome it. Even then, there is no guarantee that systematic study would actually end human ignorance about UFOs; that must await the science. But after sixty years of official denials about this potentially extraordinary phenomenon, it is time to try.

CHAPTER 28

Facing an Extreme Challenge

A deeper understanding of the unconscious aspects of the UFO taboo—the ones otherwise beyond our reach—is essential if we are to finally close the door on old ways of thinking and move this issue forward. The provocative ideas presented in the previous chapter may not answer all the questions, but the two political scientists make an intriguing and persuasive argument. They state that the fundamental problem afflicting true understanding of UFOs is ignorance, not secrecy, and that this ignorance is accepted because it serves a political purpose. Hidden forces and fears lurking under the surface of this political ignorance sustain it, while also transforming it into something far more potent: an active denial and zealous prohibition against even *considering* UFOs as a serious subject. The problem is more energized, more confrontational than simple ignorance, as we have seen. It manifests as the familiar taboo, something so accepted and taken for granted that most of us have never thought twice about it.

That political purpose is a powerful one: to maintain the imperative that we must avoid facing the possibility that *any* UFOs could be extraterrestrial. For if they were, that would mean that these miraculous craft, vehicles, objects of unknown origin—whatever they are—are generated by a more powerful "other" from somewhere else. Such a concept is simply unacceptable, and can generate a primordial terror in human beings. We take care of this through the political strategy of denying that UFOs exist at all, a stance that protects us, however temporarily, from having to confront this unthinkable threat to our core stability.

Scientists have their own reasons to be fearful. UFOs demonstrate characteristics appearing to contradict the fundamental laws of physics on which our understanding of the universe is based; if scientists did make a concerted effort to identify them, is it possible they might find the phenomenon somehow "unknowable" through our current method-ologies? So far, the UFOs have made any study difficult—they come ever so close, but not quite close enough. Does this mean we might never be able to learn what they are, even if we tried? Maybe, all of a sudden, the phenomenon will reveal itself to us before we know much of anything about it, and we'll be powerless to react.

Each of us can explore the roots of our own resistance to accepting the reality of UFOs, a process that hopefully has already begun for most readers. We may not be fully aware of buried responses and thought pat-terns, especially since the resistance is universally accepted. When they ridicule UFOs, skeptics do not consciously worry about abstractions such as anthropocentric humanism, or the loss of statehood, or the threat of annihilation, but that doesn't mean these issues do not underlie their knee-jerk reactions. Government officials don't actively contemplate such fears either, when choosing to ignore UFOs or to keep information from the public, following the decades-old trend. Scientists conveniently claim there is no evidence, but they are not thinking about the potential challenge UFOs bring to the foundation of science as they know it. So much operates outside our field of conscious awareness, perpetuating a kind of blindness.

A personal exploration might reveal only a strange discomfort with the whole notion of UFOs, an automatic, instinctual avoidance of the challenge they inherently represent. As Wendt and Duvall describe it, "the UFO taboo is akin to denial in psychoanalysis." Without pondering it, many would probably say they can't put their finger on what this chal-lenge really is. For those willing to examine further, perhaps the "skepti-cal arguments" articulated in the previous chapter will surface; or, for others, there will be religious conflicts. Most of us would prefer not to contemplate the subject at all, because we have been handed a conve-nient way out—an accepted prohibition against "believing in UFOs" that allows us to identify with the "elite" position. My hope is that, maybe now, having digested all the material presented in this book, those who have managed to come this far will not be as easily influenced by this transpar-ent taboo as they were before.

Unconscious fears about the implications of UFOs most likely lodged in the larger mind of the American political system beginning in the late 1940s, when UFOs first burst upon the scene at a national level. Yet a certain portion of the American population was already predisposed to view reports of "flying saucers" as hoaxes or exaggerations. In 1938, Orson Welles's famous radio broadcast of *The War of the Worlds* panicked numerous listeners with its all-too-realistic dramatization of an invasion by Martian spaceships, presented as if it were a live, unfolding news report. People actually fled their New Jersey homes—the site of the alleged invasion—and many others were convinced that the Earth was indeed under attack and we all would die. The broadcast tapped into an entirely different kind of fear than Americans had ever encountered before, something inexplicably terrifying. Those impacted by this would have a harder time trusting future reports of unidentified flying objects, and in this sense, a self-imposed discomfort with UFO reports was reinforced at the very outset.

But in those early years and into the 1950s, we were in our infancy when dealing with the possible meanings of the UFO phenomenon. Military and intelligence agencies were preoccupied with the task of trying to discern what these things might be in the context of the Cold War. The U.S. Air Force coped with public concerns by trying its best to explain away all UFOs, and if it couldn't, by pretending that it could. This incipient denial, bolstered by the 1953 Robertson Panel and then strengthened by the 1968 Condon report, has become even more entrenched over time. Perhaps as we learned more about UFOs after the close of Project Blue Book, gaining a clearer picture of at least their characteristics and behavior, we progressively had more reason to be worried about their threatening aspects. When J. Allen Hynek battled the problem of the taboo in the 1980s, he noted that officials had "a powerful desire to do nothing." But he also added ominously that "history has shown that in time the dam breaks, sometimes cataclysmically."

At this point, we have the option of encouraging the dam to break— slowly and methodically, rather than cataclysmically, if possible. We must recognize that the potential dangers of acknowledging and investigating UFOs are real. The fears are understandable, and even justified; and yes, the repercussions could be socially destabilizing.

But no matter how this enigma is eventually resolved, the American political establishment is monopolizing any decision making for the time being. Official bodies within other countries have obviously not been overcome by projected fears, nor do they think that any risks inherent in discovery justify ignoring UFOs. They are already moving forward, and I suspect most of these officials believe it is more dangerous to ignore UFOs than it is to confront them. The majority of the American public, as shown by various polls, already recognize the reality of UFOs, and they don't appear to be traumatized about it. Rather, they seem to want to know more.

For the benefit of the political establishment, I believe that bringing any and all fears to consciousness is our only choice. When we decide, as a society, to honestly deal with UFOs, we will be entering into a large-scale "therapeutic" process that will diminish, or even ultimately extinguish, the power of the forces sustaining the taboo. By finally shedding light on these dynamics, we will disarm them. This is perhaps the *only* way for all of us to take the next step, because it will undermine the very foundation of the dysfunctional political system in place, the central obstacle standing in our way.

In the meantime, I hope all the writers for this book have helped assuage some of that existential anxiety. Understanding brings relief, and, as the clichés say, knowledge is power and the truth will set you free. As true "militant agnostics," we can recognize that political change must incorporate these more philosophical considerations. As in Hynek's metaphor, the waters are rising to a level that will eventually compel the dam to break. We *can* find a healthy resolution to the challenge of UFOs and all they represent, and we *must* do so.

With the launching of a new U.S. government agency and the liberation of new resources, science could take its rightful place in the study of UFOs by claiming the subject as its own and beginning a new inquiry. Such a scenario would represent a dramatic turnaround from a past in which a few noble scientists made an effort to bring this controversial issue to the table, while others, although interested, were inhibited by the risk of professional ridicule. The rest succumbed to the notion that there was nothing there worth studying, as put forth in the summary of the Condon report.

A few scientists have actively studied and investigated UFOs despite the professional obstacles, and we have much to learn from them despite the passage of time. In 1968, the House Science and Astronautics Committee heard the testimony of Dr. James E. McDonald, senior atmospheric physicist of the Institute of Atmospheric Physics at the University of Arizona and a member of the National Academy of Sciences, who had spent two years investigating UFO cases. As a result of his focused study—a rarity within his profession—McDonald told the congressional committee that "no other problem within your jurisdiction is of comparable scientific and national importance," and this extraordinary matter should not be ignored. If other scientists had bothered to undertake such studies, many would have reached the same conclusion, and we'd be in a very different situation today. Instead, shortly thereafter, the University of Colorado's biased and misleading report quashed the efforts of pioneer scientists such as McDonald to interest the scientific community in studying UFOs.

Since then, Dr. Peter A. Sturrock, emeritus professor of applied physics at Stanford University and emeritus director of Stanford's Center for Space Science and Astrophysics, has taken the lead in combating the effects of the Condon report. In 1975, he conducted a survey of the American Astronomical Society and found that 75 percent of the respondents wished to see more information on the UFO subject published in scientific journals. Due to the fact that these journals rejected papers on UFOs and other anomalies out of hand, Sturrock founded the Society for Scientific Exploration and its *Journal of Scientific Exploration,* which began publication in 1987.

Sturrock is perhaps one of the most eminent scientists ever to apply the conventional scientific method to the UFO phenomenon. He has received awards from the American Astronomical Society, the American Institute of Aeronautics and Astronautics, Cambridge University, the Gravity Foundation, and the National Academy of Sciences. The American Institute of Aeronautics and Astronautics noted his "major contribution to the fields of geophysics, solar physics and astrophysics, leadership in the space science community, and dedication to the pursuit of knowledge." He has published five edited volumes, three monographs, three hundred articles and reports, and a 2009 memoir.

In 1997, Sturrock initiated and directed the first major scientific inquiry into the UFO phenomenon since the Condon study, in order to see what a new group of scientists would conclude about UFOs. A four-day conference was convened in upstate New York to rigorously review physical evidence associated with UFO reports. Seven investigators—including

Jean-Jacques Velasco and Dr. Richard Haines—presented well-researched cases with photographic evidence, ground traces and injuries to vegetation, analysis of debris from UFOs, radar evidence, interference with automobile functioning and aircraft equipment, apparent gravitational or inertial effects, and physiological effects on witnesses. The review panel of nine scientists from diverse fields—most were "decidedly skeptical agnostics" who did not have prior involvement with UFOs, according to Sturrock—reviewed the presentations and provided a sober, carefully worded summary. Although they were unable to conclude anything specific in such a short time, the panel recommended continued careful evaluation of UFO reports. It recognized that the Condon study was out of date, and that whenever there are unexplained phenomena, of course they should be investigated. And yes, the further investigation and study of UFO data *could* contribute to the resolution of the UFO problem. Those remarks were a significant advance on the position of the scientific establishment.

Still, this review didn't change much. Scientists continue to face obstacles, Sturrock notes, such as: a lack of funding for research, a false assumption that there is no data or evidence, the perception that the topic is "not respectable," and the a priori rejection of research papers by journals. One impediment is that instead of looking at the data and taking steps to acquire more, many scientists have tended to interpret the issue theoretically and then give a theoretical reason for dismissing it. For example, Astronomer Frank Drake stated in 1998 that if UFO reports are real, they must be due to extraterrestrial spacecraft. However, interstellar travel is impossible, therefore the reports must be discounted. This argument boils down to the familiar skeptical assertion that it cannot happen, therefore it does not happen. "In normal scientific research, observational evidence takes precedence over theory," Sturrock points out. "If it does happen, it can happen."

In January 2010, the prestigious Royal Society of London convened a two-day conference on "the detection of extraterrestrial life and the consequences for science and society." Physicists, chemists, biologists, astronomers, anthropologists, and theologians came together—along with representatives from NASA, the European Space Agency, and the UN Office for Outer Space Affairs—to discuss the scientific search for extraterrestrial intelligence. But one issue was not part of the mix: the still unexplained UFO phenomenon. Once again, it was as if the whole mass of evidence simply doesn't exist. And I am quite sure that if any presenters were open or curious, perhaps even informed, about the subject, they would never risk saying so among such esteemed colleagues at a high-profile forum. But the fact that this meeting took place at all, and received international media coverage, illustrates the increasing fascination and greater acceptance being afforded the search for life beyond planet Earth. I believe that after the United States establishes its own government agency to spur UFO research, and thereby changes attitudes within the scientific community, the next such conference will include a credentialed speaker on the mystery of UFOs.

Gradually, science will sort out the wheat from the chaff, and devise a way to integrate the so far unorganized UFO data into its own framework. Specific steps to be taken have been suggested by some concerned scientists, but lie outside the scope of this book. However, radical changes to the accepted scientific norm—anything leading to profound shifts in understanding—have never come about easily. UFOs seem to be the first to challenge something as fundamental as our anthropocentric, or human-centered, worldview, which could mean that resistance to studying them may turn out to be the longest in human history.

As defined by the philosopher of science Thomas S. Kuhn, author of the classic 1962 study *The Structure of Scientific Revolutions,* the process of a paradigm shift begins when a persistent anomaly is discovered that can't be explained by the existing set of assumptions within the current scientific framework. The unexplained phenomenon undermines the foundational tenets of the prevailing worldview. When the anomaly first shows itself, its implications and physical characteristics seem absolutely inconceivable, totally outside the boundary of what could be real, thereby requiring dismissal by the establishment. At first its presence is rejected as an error and often ridiculed, with proponents of its legitimacy scorned and persecuted, their jobs and reputations at risk. As evidence mounts and it can no longer be discarded, attempts are made to incorporate it and define it within the parameters of the existing paradigm. The threat to current understanding is heightened and the establishment clings ever tighter to its self-defining, and self-defined, reality, as if confronted with death. At the same time, as Kuhn describes it, the old paradigm boundaries begin to soften, and a few highly placed scientists start exploring the study of the anomaly, gradually attracting additional researchers into the fold. Finally, the new reality breaks through, often suddenly and quickly, sometimes precipitated by the efforts of a single scientist acting at a crucial time. The anomaly then becomes part of the expected and we're able to see nature in a new way, and soon the once-radical discovery becomes part of the known.

Kuhn writes: "A scientific revolution is a noncumulative developmental episode in which an older paradigm is replaced in whole or in part by an incompatible new one . . . the normal-scientific tradition that emerges from a scientific revolution is not only incompatible but often actually incommensurable with that which has gone before."

With regard to the anomaly of the UFO, it's easy to recognize its potential to create a "paradigm shift," depending on what is discovered once science decides to recognize it. Because of the extraterrestrial possibility—a challenge to our understanding of the physical universe and our place in it—there is, indeed, a risk of a very large scientific revolution. If the UFO is determined to be a secret technological creation of mankind or something more complex such as a manifestation of nature from perhaps another dimension, the discovery would be potentially transformative. And Kuhn says it can all happen due to one defining, "noncumulative" event—perhaps one pivotal, lengthy UFO display, a new type of explosive physical evidence, or even communication via radio waves or other more advanced means—an event that will leave scientists certain as to the nature and origin of the phenomenon.

Unfortunately, history shows that such change usually progresses slowly in the buildup to that defining moment. Based on scientific observations in the early sixteenth century, Copernicus proposed the heliocentric model, according to which the Earth was not stationary at the center of the universe, as orthodox science claimed, but in fact was spinning on its axis, and the planets were moving around the sun rather than the Earth. The movements of the planets were anomalies at the time, and couldn't be explained within the accepted model. Copernicus acquired data that supported this new theory and explained the observed anomalies. But, despite his rationality, his findings were considered impossible—it can't be, therefore it isn't—given what was then understood to be true. Worse, as we human beings gazed out to space in a state of ignorance, secure on our fixed planet Earth, his theory also defied our self-imposed religious dogma. A hundred and fifty years passed before the fact that the Earth revolves around the sun was accepted, and only after Galileo, Kepler, and Newton contributed in turn. Finally, humanity witnessed the emergence of the new scientific paradigm. It had been a long and painful road. Galileo had been forced by the church to retract his ideas, and was placed under house arrest for maintaining what was actually the correct view.

Smaller discoveries, even though they, too, are initially considered impossible, can shift the norm more expediently. In the early nineteenth century, scientists rejected the idea that rocks could fall from the sky, despite reports to the contrary by multiple eyewitnesses. The consensus was that this couldn't possibly be, so anyone who said otherwise must be lying, crazy, or a hoaxer. Finally, a scientist collected meteorite fragments reported by villagers in France, which were then studied in the lab, proving the reality of rocks from the sky, and the new phenomenon of meteorites was accepted from that moment on.

Presently, a few physicists are beginning to put forward theories that could explain faster-than-light travel through space, including concepts such as space travel through wormholes, multiple dimensions, and even time travelers. According to an August 2009 *Newsweek* cover story, scientists now estimate that 100 billion suns in the Milky Way galaxy support Earth-like planets in orbit around them. Given how many stars there are and the number of extrasolar planets already discovered, the chance of life existing elsewhere in the universe is very high. NASA's Kepler spacecraft was launched in 2009 to hunt for some of these planets among 100,000 stars in the constellations of Cygnus and Lyra, with the hopes of finding some terrestrial planets with habitable conditions. As of this writing, we've already found over 400 planets orbiting other stars. By 2013, Kepler is likely to have located hundreds, if not thousands, of potentially habitable planets. NASA has also developed a highly sensitive infrared space telescope now searching for small, dark asteroids and other near-Earth objects in our solar system, and it sent its first images back through space in January 2010.

Through its persistent recurrence, the UFO phenomenon makes its own demands on scientists, who should no longer be allowed the luxury of denial. We have always been an evolving species seeking to understand the unknown, and we will handle whatever changes come from radical new discoveries. As Kuhn said many years ago, "when paradigms change, the world itself changes with them."

Over the years, debunking organizations have developed the slogan "Extraordinary claims demand extraordinary evidence" as a kind of mantra, rolling all their objections into one, which is used to dismiss UFOs out of hand. They're claiming that there is not sufficient evidence to support the "claim" that UFOs exist.

This book has accomplished, in my view, the presentation of some of the very compelling evidence—only a slice of it, we must remember—that UFOs *do* exist. We have seen that there are solid, three-dimensional objects of unknown origin flying in our skies, stopping in midair and zooming toward outer space, which are apparently not natural or man-made. They've come very close and landed as well, leaving physical traces in soil while shriveling the leaves of nearby plants. They interact with aircraft and have physical effects upon them. Photographs have captured their image on film, and radar blips have done the same on tracking monitors. Thousands of people from all walks of life in every continent have seen these objects, including many pilots and military officers. The group represented in this book, myself included, understand that what the skeptics love to call a "claim"—the existence of unknown objects in the sky—is actually an established fact. There is more than enough evidence to determine that *something physical* is there.

We in this group are also "militant agnostics": we don't know what this something is, nor do we know what it is *not*. We are not making an extraordinary claim, because we're not claiming anything beyond the reality of a physical phenomenon, and the five premises that stem from this reality as outlined in the introduction to this book. Yes, that phenomenon is definitely extraordinary. The basic misunderstanding underlying the skeptics' catchy buzz-phrase—"Extraordinary claims demand extraordinary evidence"—is, once again, the equating of UFOs to extraterrestrial spacecraft by definition. When the debunkers rally around this battle cry and dismiss all the evidence with a wave of their hand, this is really what's on their minds; otherwise, there would be no need for them to be so blindly defensive, and even hostile.

Their concern is understandable, even if it's dealt with dishonestly.

The COMETA group pointed out at the very beginning of this journey, and many of our contributors have stated as well, that the extraterrestrial hypothesis is the most likely one to explain what we know. That's a very loaded proposition, but we're stuck with it. And actually, it is not an extreme position, in comparison to the two polarized positions that are so common in the culture: either we know already what UFOs are (alien spacecraft), or they can't possibly exist at all, and therefore don't. These two extremes are the *real* extraordinary claims.

We ask those on the two sides of this outmoded contest between unwavering believers and nonbelievers to realize the fallacy of both positions, and to accept the logic, necessity, and realism of the agnostic view. Scientists must disavow the untenable claim that we have no evidence other than eyewitness reports, which are to them—of course—unreliable. That is another "extraordinary claim" that doesn't hold up, as this book attests.

The time has come to proceed logically. Given that we know we have a physical manifestation of something highly unusual of unknown origin, isn't it time to acquire the additional evidence needed to find out what it is? If we need extraordinary evidence, then let's do our job and go get it. We Americans will have the cooperation of other scientists from around the world who have already invested their limited resources into such an endeavor. And so a new slogan is in order: "An extraordinary phenomenon demands an extraordinary investigation." The world's scientists are entirely capable of devising the methodologies and manufacturing the technology needed to solve this extraordinary mystery.

As the contributors here have shown, there is too much at stake to continue stonewalling. At the same time, we can't deny the fact that there is a risk in moving forward. The phenomenon itself has placed us in a precarious situation that we have not chosen, and that we can do nothing about. We must strive to learn what we can, for it's in our deepest nature and best interest to do so—to simply want to find out. Perhaps this discovery will be a turning point in our history. Perhaps not. But most likely, there's something supremely important locked up in the UFO phenomenon that could be transformative for all of us. It's time now, finally, to open our eyes and find out what that might be.

ACKNOWLEDGMENTS

First and foremost, I'd like to thank the eighteen distinguished contributors whose pieces form the essential core of this book, and who made it all possible. It has been a privilege to work with this exceptional group. My deepest thanks to each of them for their trust and for their diligent work on many drafts. These men have courageously gone on the record about the reality of UFOs, and I hope others in comparable positions will now be moved to do the same.

I extend a special appreciation to John Podesta for his eloquent foreword and for his ongoing public support of the Coalition for Freedom of Information (CFi). His brilliance and honesty are inspiring. Others made major contributions to the text: Yves Sillard of GEIPAN wrote an important commentary, and André Amond, J. Dori Callahan, Julio Chamorro, Anthony Choy, Jean-Pierre Fartek, Will Miller, and Robert Salas provided interviews and helpful material. I am very grateful to former Arizona governor Fife Symington III for his encouragement, which helped launch the book.

Phyllis Wender, my agent from the Gersh Agency in New York, believed in this project from the outset. I thank her wholeheartedly for her appreciation of the approach embodied by the book and her unwavering determination to see it published. Her wise advice has been indispensable, and her assistant Lynn Hyde also deserves my thanks. At the Crown Publishing Group, I am indebted to Shaye Areheart for her vision, leadership, and commitment to the book, and to my enthusiastic editor, Kate Kennedy, who guided me throughout the lengthy publishing process and made many significant editorial contributions that improved the manuscript.

A special thanks goes to my close friend Budd Hopkins for providing daily, steady support as I dealt with the myriad personal and professional challenges inherent in producing this book. He dutifully read and reread every word of the manuscript at its various inceptions and offered many perceptive edits and suggestions. I'm also grateful to David M. Jacobs, Paul McKim, and Lloyd Garrison for reading parts of the manuscript and providing useful feedback.

I cannot overlook two key colleagues who profoundly influenced my life before I was unexpectedly confronted with the issue of UFOs. Burma activist and writer Alan Clements inspired me with his compassionate activism and commitment to a people's struggle, and opened up a new world to me. Investigative reporter Dennis Bernstein, host of *Flashpoints* on Pacifica radio, taught me the principles and craft of advocacy journalism, leading me into the world of freelance publishing and, eventually, radio broadcasting. I can't thank my dear friends Alan and Dennis enough for giving me the foundation that made it possible for me to later take on the risky subject of UFOs.

At the beginning of my UFO explorations, Ralph Steiner helped me navigate and offered much reassurance, Stephen Bassett was supportive, and Clifford Stone, Steven Greer, and Grant Cameron provided me with hundreds of government documents released through the Freedom of Information Act. I thank *Boston Globe* editor Chris Chinlund and Robert Whitcomb from the *Providence Journal* for publishing my first UFO articles.

I am very grateful to Larry Landsman, my partner in the CFi, for opening so many doors for me, and for his consistent advice and comradeship over all these years. Without Larry, this book would never have been born. I also appreciate the invaluable education provided by Ed Rothschild, senior public affairs strategist with the Podesta Group. And I extend my thanks to James Fox, Stan Gordon, Lee Helfrich, and Jeff Sagansky and the team at Break Thru Films for the meaningful opportunities they provided me.

Many skilled investigators have spent decades collecting data on UFOs, and I relied on their work continuously throughout the book. I pay special tribute to veteran researcher Richard Hall, who died of cancer in 2009, and who was always available to answer my questions. Along with others already mentioned, I'm also personally indebted to researchers Jerome Clark, Peter Davenport, Richard Dolan, Stanton Friedman, A. J. Gevaerd, Timothy Good, Bernard Haisch, Bruce Maccabee, Mark Rodeghier, Ted Roe, Brad Sparks, Peter Sturrock, Rob Swiatek, and Nancy Talbott.

Pituka Heilbron and Andrea Soares Berrios spent much time translating both text and many e-mails. Thanks also to Jean-Luc Rivera and Oscar Zambrano for translations, and to Jean-Claude Ribes, Valery Uvarov, Ruben Uriarte, and André Morin. Others assisted with various aspects of the book: Yvan Blanc, Joaquim Fernandes, Kelly Fox, Seth Keal, Phil Imbrogno, Charles Miller, Gustavo Rodríguez, Susan Stanley, and Bernard Thouanel; and at Crown, thanks to Mark Birkey, Jill Browning, Lenny Henderson, Kyle Kolker, Elizabeth Rendfleisch, Kira Walton, and Campbell Wharton.

Finally, I thank my mother, Ellen S. Kean, and my father, Hamilton F. Kean, for their steadfast, unconditional support and genuine enthusiasm for this project, despite the taboo nature of its subject matter. Thanks for having faith in me.

ABOUT THE CONTRIBUTORS

RAY BOWYER has been a flight calibration inspection pilot, and continues as a commercially qualified airline pilot. He has flown for ten airlines operating in Europe and the Middle East, including Jersey European, Channel Express, Regionair, BusinessAir, and Farner Air. From 1999 to 2008, he was a Line Captain for Aurigny Air Services in the Channel Islands, flying inter-island and international routes based in Guernsey. He currently flies as a captain for a Channel Island–based corporation throughout Europe and has a total flying time of 7,000 hours.

WILFRIED DE BROUWER spent twenty years as a fighter pilot in the Belgian Air Force. He was then appointed to the Strategic Planning Branch of NATO in 1983, while a Colonel. After that, he became Wing Commander of the Belgian Air Force Transport Wing and, in 1989, chief of the Operations Division in the Air Staff. Promoted to Major General in 1991, De Brouwer served as Deputy Chief of Staff of the Belgian Air Force. Beginning in 1995, after retiring from the Air Force, he worked for more than ten years as a consultant for the United Nations to improve the UN Logistics rapid-response capabilities during emergencies.

JOHN J. CALLAHAN has over thirty years of experience at the Federal Aviation Administration (FAA) specializing in the air traffic control centers. As Automation Branch Chief, he supervised the design, programming, testing, and implementation of all air traffic control facilities software programs. From 1981 to 1988, he was Division Chief for Accidents, Evaluations, and Investigations at Washington Headquarters, where he was responsible for the quality of air traffic service provided to FAA users. After retiring, Callahan was employed as a Senior Analyst for Washington Consulting Group and Chief Executive Officer for Crown Communications Consulting Company. He now owns and operates Liberty Tax Service in Culpeper, Virginia.

RAYMOND DUVALL is Morse-Alumni Professor and Chair of the Department of Political Science at the University of Minnesota. His co-edited publications include *Power in Global Governance* (Cambridge University Press, 2005) and *Cultures of Insecurity: States, Communities and the Production of Danger* (University of Minnesota Press, 1999). His recent articles have appeared in such scholarly journals as *International Organization* (2005–06), *Millennium* (2007), *Review of International Studies* (2008), and *Political Theory* (2008). Dr. Duvall's teaching and research focus on facets of critical international relations theory, including the productive effects of social practices.

RODRIGO BRAVO GARRIDO is a captain and pilot for the Aviation Army of Chile. In 2000, at age twenty-four, he was assigned to conduct an internal study titled "Introduction to Anomalous Aerial Phenomenon and Their Considerations for Aerospace Security," involving previous case reports of military planes' encounters with UAP. He has since continued this research and now works in cooperation with the Committee for the Study of Anomalous Aerial Phenomena (CEFAA), a branch of the General Administration of Civil Aeronautics, Chile's equivalent of our FAA.

JÚLIO MIGUEL GUERRA became a pilot with the Portuguese Air Force in 1973 and was an operations officer specializing in accident prevention at Ota Air Base. In 1990 he began flying commercially with Air Atlantis, a charter of Portugal's national airline TAP, Air Columbus, and Air Atlanta, piloting Boeing 737-200/300 jets. Since 1997, he has been a Line Captain for Portugalia Airlines. He is also a private flight instructor and an examiner for the Joint Aviation Authorities, a European body developing and implementing common safety regulatory standards. With 18,000 hours of flight time, Captain Guerra received an Aeronautic Science Degree from Lusófona University in 2009.

RICHARD F. HAINES is a senior research scientist who worked at NASA–Ames Research Center from 1967 to 1988 on projects such as Gemini, Apollo, Skylab, and the International Space Station, and managed the Joint FAA/NASA Head-up Display Evaluation Program. He was appointed Chief of the Space Human Factors Office at NASA–Ames in 1986. Dr. Haines has published more than seventy-five papers in leading scientific journals and over twenty-five U.S. government reports for NASA. Since retiring in 1988, he worked as a senior research scientist for the Research Institute for Advanced Computer Science, RECOM Technologies, Inc., and Raytheon Corporation. Currently, he serves as Chief Scientist for the National Aviation Reporting Center on Anomalous Phenomena (NARCAP).

CHARLES I. HALT was a Lieutenant Colonel when he was assigned to RAF Bentwaters, England—the largest Tactical Fighter Wing in the U.S. Air Force—as Deputy Base Commander and then as Base Commander. After becoming a full Colonel, he was Base Commander at Kunsan Air Base in Korea, the F-16 base responsible for any offensive action required on the Korean peninsula, and also was instrumental in establishing the Cruise Missile Base in Belgium. Finally, he served as Director, Inspections Directorate, for the DoD Inspector General, with total inspection oversight of the entire Department of Defense. Colonel Halt retired in 1991 and now manages a large gated community.

OSCAR SANTA MARÍA HUERTAS was a jet fighter pilot with the Peruvian Air Force (FAP) for many years, with flying experience in T-41D, T-37, A-80, T-33, A-37, MB-399, and SU-22 aircraft. He was stationed in numerous military bases throughout Peru and was chief of the Academic Department and a flight instructor in the Officers School of the FAP. Santa María also spent eleven years in the Air Force Accident Prevention and Investigation Department. He retired with the rank of comandante (equivalent to a colonel) in 1997 but remains active, currently working as a consultant in Flight Safety and Accident Prevention for the airline industry in Peru.

PARVIZ JAFARI is a retired General of the Iranian Air Force. After joining the Air Force, he spent two years training in the United States, at Lackland Air Force Base in Texas, Craig AFB in Alabama, and Nellis AFB in Nevada. In his country, Jafari served as a base commander for several bases and an Air Force Headquarters operations officer. As a General, Jafari became the coordinating officer between the Iranian Army, Navy, and Air Force. He retired in 1989 and lives in Tehran.

DENIS LETTY is a well-known fighter pilot and Major General in the French Air Force. He was head of the Fifth Fighter Wing, Strasbourg Air

Base, French Air Force South East Defense Zone, and the French Military Mission near Allied Air Forces Central Europe. As a commander, he was decorated with the Legion of Honor. After retiring, General Letty served as president of the joint venture company Aviation Defense Service, which provided electronic warfare training for the armed forces. He also became chairman of the COMETA group, a private, in-depth fact-finding committee formed to study the UFO phenomenon, which published the report "UFOs and Defense" in 1999.

JAMES PENNISTON entered the Air Force in 1973 and was assigned to the Strategic Air Command Elite Guard in Omaha, Nebraska, working security for the SAC Command Post. Subsequent assignments took him to RAF Alconbury in England and Malmstrom Air Force Base in Montana, as a Flight Security Controller for the protection and launch readiness of Minuteman ICBMs. In 1980, he was placed in charge of Security Police Plans and Programs at RAF Bentwaters, England. Numerous other assignments followed, including service in Desert Shield and Desert Storm. He retired from the military in 1993 and now works as a Human Resource Manager for manufacturing and county government in Illinois.

JOSÉ CARLOS PEREIRA is a Brazilian four-star Brigadier General, now retired. He was a commander of several air bases in Brazil and commander of the Brazilian Air Force Academy. In 1999, he became a commander of the Brazilian Airspace Defense Command, known as COMDABRA. From 2001 to 2005, he served as General Commander of Air Force Operations, which required his supervision of thirteen generals and 27,000 subordinates. In 2006, after retiring from the Air Force, Brigadier General Pereira was appointed President of the Brazilian Airport Infrastructure Agency, the government agency responsible for airport management, from which he has now retired.

JOHN PODESTA was the White House Chief of Staff to President William J. Clinton. He also served in the president's cabinet and as a principal on the National Security Council. Most recently, he was a co-chair of President Obama's transition team, for which he coordinated the priorities of the incoming administration's agenda, oversaw the development of its policies, and spearheaded its appointments of major cabinet secretaries and political appointees. Since 2003, he has been the President and CEO of the Center for American Progress, a leading organization in the development of and advocacy for progressive policy. Podesta is the author of *The Power of Progress: How America's Progressives Can (Once Again) Save Our Economy, Our Climate, and Our Country.*

NICK POPE worked for the British Ministry of Defence for twenty-one years, from 1985 to 2006. His career involved postings to policy, operations, personnel, finance, and security divisions. During the first Gulf War he was recruited into the Joint Operations Center, where he worked in the Air Force Operations Room as a watchkeeper/briefer. From 1991 to 1994, Pope's primary duty was to investigate reports of unidentified flying objects and assess whether any sightings were of defense interest. Various promotions followed, and his last posting was to the Directorate of Defence Security. Now retired, Nick Pope works as a freelance journalist and broadcaster.

RICARDO BERMÚDEZ SANHUEZA is a retired General for the Chilean Air Force who served as Chilean Air Attaché in London and was Chief Commander of the Air Force's Southern Area. He was also Director of the Technical School of Aeronautics. In 1998, he co-founded the Committee for the Study of Anomalous Aerial Phenomena (CEFAA), a branch of the General Administration of Civil Aeronautics, Chile's FAA, to study aviation incidents involving anomalous aerial phenomena. He was appointed first President of the CEFAA and served until 2002. In January 2010, he was reinstated as Director of the CEFAA and now works full-time investigating UFO incidents involving civil or military aeronautic personnel.

FIFE SYMINGTON III was the Republican Governor of Arizona from 1991 to 1997. He was also Chairman of the Western Governors' Association. A decorated Air Force veteran of the war in Southeast Asia, Symington is a cousin of the late Stuart Symington, Democratic Senator from Missouri. After leaving office, Mr. Symington co-founded the Arizona Culinary Institute and the Symington Group, a strategic, political, and business consulting company. In 2007, he and his partners founded the Independent Energy Group of Arizona, which specializes in the development of commercial solar arrays. A long-time pilot, he frequently flies his twin-engine Beechcraft Baron plane between his two homes in Phoenix and Santa Barbara, California.

JEAN-JACQUES VELASCO was an engineer at the French National Center for Space Studies (Centre Nationale d'Études Spatiales, CNES), specializing in satellite research. In 1977, he joined a new French team formed to study unidentified aerospace phenomena within CNES. He became the director of this agency in 1983, and remained in that position until 2004, becoming an international authority on the scientific study of UFOs. His advice was sought by countries wishing to establish their own government agencies to investigate UFOs, such as Chile and Peru, and by the European parliament in 1994. He is the author of several books on the subject of UFOs.

ALEXANDER WENDT is Mershon Professor of International Security at the Ohio State University. Previously, he taught at Yale University, Dartmouth College, and the University of Chicago. He is interested in philosophical aspects of international politics, and has published a number of articles in leading political science journals, as well as a 1999 book, *Social Theory of International Politics* (Cambridge University Press), which received the International Studies Association's Best Book of the Decade Award in 2006.

ABOUT THE AUTHOR

LESLIE KEAN is an independent investigative journalist with a background in freelance writing and radio broadcasting. She has contributed articles to dozens of publications here and abroad, including the *Boston Globe, Philadelphia Inquirer, Atlanta Journal-Constitution, Providence Journal, International Herald Tribune, Globe and Mail, Sydney Morning Herald, Bangkok Post, The Nation,* and *The Journal for Scientific Exploration.* Her stories have been syndicated through Knight Ridder/Tribune, Scripps-Howard, the *New York Times* wire service, Pacific News Service, and the National Publishers Association. While spending many years reporting on Burma, she co-authored *Burma's Revolution of the Spirit: The Struggle for Democratic Freedom and Dignity* (Aperture, 1994), and she has contributed essays for a number of anthologies published between 1998 and 2009.

Kean was also a producer and on-air host for a daily investigative news program on KPFA radio, a Pacifica station. In 2002 she co-founded the Coalition for Freedom of Information (CFi), an independent alliance advocating for greater government openness on information about UFOs and for responsible coverage by the media based on a rational and credible approach. As director of the CFi, she was the plaintiff in a successful four-year Freedom of Information Act federal lawsuit against NASA. Kean was a producer for the 2009 independent documentary *I Know What I Saw* and is currently working with Break Thru Films, an award-winning film company, on a new feature documentary.

NOTES

INTRODUCTION

1 *Now known as the COMETA Report* The acronym COMETA is an abbreviation for Comité d'Études Approfondies (Committee for In-Depth Studies), the name of the committee that conducted the study.

2 *"the most logical explanation for these sightings"* The COMETA Report, "UFOs and Defense: What Should We Prepare For?" Written by the French association COMETA, 1999. "Le rapport Cometa, les Ovni et la Defense, A quoi doint-on se preparer?" G.S. Presse Communication, 1999. Editions du Rocher, 2003. Appeared in the magazine *VSD* in France, July 1999.

2 *Among them, all retired, were a four-star general* COMETA members and contributors include: General Bernard Norlain, former commander of the French Tactical Air Force; Andre Lebeau, former head of CNES; General Denis Letty of the Air Force, former auditor (FA) of IHEDN; General Bruno Lemoine of the Air Force (FA of IHEDN); Admiral Marc Merlo (FA of IHEDN); Jean-Jacques Velasco, head of SEPRA/GEPAN; Michel Algrin, doctor in political sciences, attorney at law (FA of IHEDN); General Pierre Bescond, engineer for armaments (FA of IHEDN); Denis Blancher, chief national police superintendent at the Ministry of the Interior; Christian Marchal, chief engineer of the national *corps des Mines,* research director at the National Office of Aeronautical Research (ONERA); General Alain Orszag, Ph.D. in physics, engineer for armaments. Other contributors include François Louange, president of Fleximage, specialist in photo analysis; General Joseph Domange of the Air Force.

4 *UFOs became the focus* Leslie Kean, "UFO Theorists Gain Support Abroad, but Repression at Home," *Boston Sunday Globe,* May 21, 2000.

6 *Volumes of case studies have been published* There are too many to mention, including many white papers, transcripts, magazine stories, journal articles, and books about a specific case or one particular aspect of UFO research. Much outstanding work is also published on a number of credible websites, and other books have been written more recently. The following works cover the UFO topic in general and were of particular importance to me personally during my first few years of study, from 1999 to 2001: Edward J. Ruppelt, *The Report on Uniden-*

tified Flying Objects (Doubleday, 1956; revised edition 1959); Richard H. Hall, editor, *The UFO Evidence* (NICAP, 1964); Edward U. Condon, *Scientific Study of Unidentified Flying Objects* (Bantam Books, 1969); J. Allen Hynek, *The UFO Experience: A Scientific Inquiry* (Marlowe & Company, 1972); David Jacobs, *The UFO Controversy in America* (Indiana University Press, 1975); Lawrence Fawcett and Barry J. Greenwood, *Clear Intent* (Prentice-Hall, 1984); Timothy Good, *Above Top Secret* (William Morrow, 1988); Don Berliner, *UFO Briefing Document* (Dell, 1995); Budd Hopkins, *Witnessed* (Pocket Books, Simon & Schuster, 1996); Stanton T. Friedman, *Top Secret/Majic* (Marlowe & Co., 1996); Clifford E. Stone, *UFOs Are Real* (SPI Books, 1997); Jerome Clark, *The UFO Encyclopedia,* 2nd edition, vols. 1 and 2 (Omnigraphics, Inc, 1998); Peter A. Sturrock, *The UFO Enigma: A New Review of the Physical Evidence* (Warner Books, 1999); Richard M. Doland, *UFOs and the National Security State* (Keyhole Publishing Company, 2000); Terry Hansen, *The Missing Times* (Xlibris, 2000); Bruce Maccabee, *UFO/FBI Connection* (Llewellyn Publications, 2000); Richard H. Hall, *The UFO Evidence: A Thirty-Year Report,* vol. 2 (The Scarecrow Press, 2001). More comprehensive reading lists can be found at http://www.cufon.org/cufon/rlist/a-n.htm and http://www.cufos.org/books.html.

6 *through any one short news piece* Some examples of my additional stories are: "Pilot Encounters with UFOs: New Study Challenges Secrecy and Denial," *Providence Journal* and Knight Ridder wire service, May 3, 2001; "Open UFO Files to Rest of Us Earthlings," *Atlanta Journal-Constitution* and Knight Ridder/Tribune wire service, December 13, 2002; "Forty Years of Secrecy: NASA, the Military and the 1965 Kecksburg Crash," *International UFO Reporter* (IUR), the journal of the J. Allen Hynek Center for UFO Studies, vol. 30, no 1, October 2005; "Just What Was That Object Hovering Overhead at O'Hare?" Scripps-Howard News Service, February 26, 2007; "Former Arizona Governor Now Admits Seeing UFO," *Arizona Daily Courier,* March 18, 2007. See www.freedomofinfo.org for more about my work.

11 *"a common-sense identification, if one is possible"* Richard Haines, *Observing UFOs: An Investigative Handbook,* (Nelson-Hall, 1980), chapter 2.

CHAPTER 1: MAJESTIC CRAFT WITH POWERFUL BEAMING SPOTLIGHTS

21 *"no USAF stealth aircraft were operating in the Ardennes area"* Joint Staff, Washington, D.C., Information Report #5049, "Belgium and the UFO Issue," March 30, 1990.

21 *"there has never been any sort of American aerial test flight"* Don Berliner, *UFO Briefing Document* (Dell Publishing/Random House, 1995), p. 144.

23 a *Belgian movie producer and two colleagues* Marie-Thérèse de Brosses, from an

interview with Professor Auguste Meessen, "An Unidentified Flying Object on the Radar of an F-16," *Paris Match*, July 5, 1990.

CHAPTER 2: THE UAP WAVE OVER BELGIUM

25 *and is subsequently retrieved by the diver* The study "Étude Approfondie et Discussion de Certaines Observations du 29 Novembre 1989" by Professeur Auguste Meessen, *Inforespace*, no. 95, octobre 1997, pp. 16–70, includes descriptions of the "red ball show" at Lake Gileppe. http://www.meessen.net/AMeessen/Gileppe .pdf. These observations were also described in the first book of SOBEPS.

30 *"the lines of force" in a magnetic field* André Marion, "Nouvelle Analyse de la Diapositive de Petit-Rechain" (A New Analysis of the Petit-Rechain Slide), Orsay, January 17, 2002.

30 *as suggested by Professor Auguste Meessen* Auguste Meessen, professeur emeritus at the University of Louvain, "Réflexions sur la propulsion des Ovnis" (Reflections on UFO Propulsion), http://www.meessen.net/AMeessen/ReflexionPropulsion .pdf.

30 *such an effect would not occur if the picture was a hoax* Translated text of Professor Marion: "It seems difficult to envisage a hoax created with a model or other similar device. This is confirmed by the digital analysis (see further) . . . The existence of the 'lines of force' is a strong argument against the thesis of a hoax, which would be particularly sophisticated. Moreover, it is unclear why a forger would have bothered to imagine and realize a complex phenomenon, especially since it is not noticeable without sophisticated processing of the slide." Marion, ibid.

CHAPTER 3: PILOTS

42 *case summaries involving pilots and their crews* Richard F. Haines, "Aviation Safety in America—A Previously Neglected Factor," NARCAP Technical Report 01–2000, October 15, 2000, http://www.narcap.org/reports/001/narcap.TR1.AvSafety .pdf. I reported on this in "Pilot Encounters with UFOs: New Study Challenges Secrecy and Denial," *Providence Journal*/Knight Ridder, May 3, 2001.

42 *unidentified aerial phenomena, or UAP* See the Introduction to this book, p. 11, for Dr. Haines's definition of the term UAP.

42 *Haines said in a 2009 interview* Interview with David Biedny and Gene Steinberg for "The Paracast," April 5, 2009, http://www.theparacast.com/show-archives/.

43 *the National Aviation Reporting Center on Anomalous Phenomena* Visit www.narcap .org for more information.

45 *Neil Daniels, a United Airlines captain for thirty-five years* I met with Daniels at his home outside San Francisco and conducted follow-up interviews by phone.

Chapter 4: Circled by a UFO

47 *a captain with Portugália Airlines* Captain Guerra has 17,000 hours of flight experience, and in 2009 he received an aeronautic science degree from the Lusofona University of Oporto.

50 *a report on this incident to Project Blue Book* General Ferreira's report of September 4, 1957, is available through the Project Blue Book archives. His description bears an uncanny resemblance to that provided by General Parviz Jafari of Iran about the object he was sent to pursue over Tehran, also as an air force pilot, in 1976. Jafari presents his case in chapter 9. The details of Jafari's encounter were filed by the Defense Intelligence Agency, after the close of Project Blue Book.

Through an intermediary, I asked General Ferreira if he would speak with me, hoping this would develop into an extensive interview. Being of poor health, he declined. In 1975, Ferreira said publicly, "I think these events should be introduced and studied at the universities, because these kind of phenomena are very far from our present technological performances." It was therefore not surprising—and fortunate for Guerra and his fellow pilots—that as air force Chief of Staff, he released the data to the scientific team from various universities to conduct the study.

50 *They concluded that the object remained unidentified.* The Portugese study "UFO Daylight Report by Three Portuguese Air Force Pilots, Ota, Portugal," by the National Center for UFO Phenomenon Investigation (CNIFO), has not been translated into English. A summary of the results by J. Sottomayor and A. Rodrigues was published in *Flying Saucer Review,* vol. 32, no. 2 (1987), pp. 12–13. Now, the Center for Transdisciplinary Study on Consciousness (CTEC), an interdisciplinary academic group at University Fernando Pessoa, has assembled all the files on the UFO phenomena in Portugal, according to its cofounder, Dr. Joaquim Fernandes. For more information, e-mail ctec@ufp.edu.pt.

Chapter 5: Unidentified Aerial Phenomena and Aviation Safety

53 *will not be in error by more than an order of magnitude* Richard Haines and Courtney Flatau, *Night Flying* (McGraw-Hill School Education Group, 1992).

53 *one of the few official statements to this effect on record* U.S. Air Force Project Blue Book file WDO-INT 11-WC23, 1958.

54 *"as if by a single command"* Aerial Phenomenon Research Organization Bulletin, January–February 1969, pp. 1, 4.

57 *and interviewed Ken Hansen* This name is a pseudonym.

58 *remains even more of a mystery* Richard F. Haines and Paul Norman, "Valentich Disappearance: New Evidence and a New Conclusion," *Journal of Scientific Exploration,* vol. 14, no. 1 (2000), pp. 19–33.

59 *but this aircraft suffered no ill effects* Bruce Maccabee, "A History of the New Zealand Sightings of December 31, 1978," 2005, http://brumac.8k.com; Bruce Maccabee, "Atmosphere or UFO? A Response to the 1997 SSE Review Panel Report," *Journal of Scientific Exploration,* vol. 13, no. 3 (1999), pp. 421–59.

59 *a dive to avoid a collision* Richard F. Haines, *International UFO Reporter,* vol. 32, no. 3 (July 2009), pp. 9–18.

61 *he believed the thing was a "spaceship"* Richard F. Haines, "Commercial Jet Crew Sights Unidentified Object—Part I," *Flying Saucer Review,* vol. 27, no. 4 (January 1982), pp. 3–6; Richard F. Haines, "Commercial Jet Crew Sights Unidentified Object—Part II," *Flying Saucer Review,* vol. 27, no. 5 (March 1982), pp. 2–8.

61 *"in the Big Bateau Bay in Spanish Fort, Alabama"* NTSB Report ATL03FA008.

61 *would have produced wing-tip vortex turbulence* R. D. Boyd, "The Last Flight of Nightship 282." In preparation, 2010.

62 *the investigation conducted by the NTSB* Boyd, Ibid.

62 *"possible presence of inorganic silicate compounds"* NTSB Accident Report ATL03FA008, p. 4, undated.

62 *a recently unclassified report from the United Kingdom* Defence Intelligence Analysis Staff, Project Condign, 2000.

62 *high-quality foreign pilot reports as well* Dominique F. Weinstein, "Unidentified Aerial Phenomena: Eighty Years of Pilot Sightings," National Aviation Reporting Center on Anomalous Phenomena (www.narcap.org), Technical Report 4, 2001.

62 *intelligence and deliberate flight control* Richard F. Haines, "Aviation Safety in America—A Previously Neglected Factor," NARCAP Technical Report 01, 2000.

64 *"Nobody knows what to do, really"* National UFO Reporting Center, August 5, 1992.

CHAPTER 6: INCURSION AT O'HARE AIRPORT

65 *the* Chicago Tribune *on January 1, 2007* Jon Hilkevitch, "In the Sky! A Bird? A Plane? A . . . UFO?" *Chicago Tribune,* January 1, 2007.

69 *and five other specialists* Haines et al., "Report of an Unidentified Aerial Phenomenon and Its Safety Implications at O'Hare International Airport on November 7, 2006," March 9, 2007, NARCAP Technical Report 10, http://www.narcap.org/reports/010/TR10_Case_18a.pdf.

69 *"UAP as non-existent"* Ibid., p. 100.

70 *"a future incident such as this"* Ibid., p. 5.

71 *"whether acknowledged or unacknowledged"* Ibid., p. 54.

Chapter 7: Gigantic UFOs over the English Channel

78 *and many other avenues of investigation* Jean-François Baure, David Clarke, Paul
Fuller, and Martin Shough, "Report on Aerial Phenomena Observed Near
the Channel Islands, UK, April 23 2007," February 2008 http://www.guernsey
.uk-ufo.org/.

Chapter 8: UFOs as Air Force Targets

83 *General Jafari and Comandante Santa María* Comandante is the rank equivalent to
colonel in the U.S. Air Force.

Chapter 9: Dogfight over Tehran

87 *It was flashing with intense red, green, orange, and blue lights* Jafari's description of
the UFO, at very close range, is unusual. However, it bears an extraordinary
resemblance to a report filed by another general, when he, like Jafari, was also
an Air Force pilot. As referenced in chapter 4 by Julio Guerra, and in my note
for that chapter, Portuguese General José Lemos Ferreira submitted his description of a UFO to the U.S. Air Force's Project Blue Book in 1957. The document
is available in those archives.

 While on a nighttime practice flight with three other Air Force jets, Ferreira saw an object that looked like "a bright star unusually big and scintillating, with a colored nucleus which changed color constantly—deep green,
blue, reddish and yellowish hues." Note the similarity to Jafari's description:
"It looked similar to a star, but bigger and brighter," and then," it was flashing
with intense red, green, orange and blue light so bright that I was not able to
see its body . . . The sequence of flashes was extremely fast, like a strobe light."
The next phase is chillingly consistent in the two encounters. Ferreira says that
the pilots saw "first one small circle of yellow light coming out of the larger
object, then three others," and that these were considerably smaller than the
scintillating, main object. Jafari states later in this chapter that he saw "a round
object" leave the larger object and head toward him, looking like "a brightly
lit moon coming out over the horizon." And he, too, witnessed not just one of
these round lights ejected from the brilliant one, but a series of them. Both
incidents involved multiple Air Force witnesses. Jafari's case was reported in a
U.S. Defense Intelligence Agency document in great detail, as described later
in this book.

 It is unusual enough for pilots to get such extended close views of UFOs
while in the air; for detailed reports to be filed about them; and for the primary
witness to later be promoted to the rank of general. But when the details are
so strikingly similar—even though they were seen nineteen years apart over

two different continents—it is reasonable to wonder whether the two groups of pilots were witnessing the same, or almost identical, phenomena.

90 *for an examination and more blood tests* Exposure to radiation can reduce the production and/or aggregation of blood platelets, which are essential for coagulation. Perhaps this explains Jafari's problem, but we don't know. He does not have copies of the medical records.

CHAPTER 10: CLOSE COMBAT WITH A UFO

93 *On April 11, 1980* The first draft of this piece was translated from Spanish by Andrea Soares Berrios and Oscar Zambrano, who also translated during follow-up communications and further development of the piece. I worked on the final edits with Comandante Santa María in English.

CHAPTER 11: THE ROOTS OF UFO DEBUNKING IN AMERICA

103 *assigning a security classification and code name to it* General Nathan F. Twining to Commander, Air Material Command, "AMC Opinion Concerning 'Flying Discs,' " September 23, 1947 (contained in Edwin U. Condon, project director, *Scientific Study of Unidentified Flying Objects,* 1969), pp. 894–95.

103 *and given the code name "Sign"* Directive—Major General L. C. Craigie to Commanding General Wright Field (Wright-Patterson AFB), Disposition and Security for Project Sign, December 30, 1947 (contained in Edwin U. Condon, project director, *Scientific Study of Unidentified Flying Objects,* 1969) p. 896.

104 *repeated attempts using the Freedom of Information Act* Edward J. Ruppelt, *The Report on Unidentified Flying Objects* (Doubleday & Company, 1956), pp. 62–63. Ruppelt was the first chief of Project Blue Book, from early 1951 until September 1953. David Michael Jacobs, *The UFO Controversy in America* (Indiana University Press, 1975), p. 47. Michael D. Swords, "Project Sign and the Estimate of the Situation," *Journal of UFO Studies,* n.s. 7 (2000), pp. 27–64, http://www.ufoscience.org/history/swords.pdf.

104 *"from another nation in this world"* W. P. Keay, FBI memorandum, "Flying Saucers," July 29, 1952 (contained in Bruce Maccabee, *UFO FBI Connection* (Llewellyn Publications, 2000).

104 *"seriously considering the possibility of planetary ships"* W. P. Keay, FBI memorandum, "Flying Saucers," October 27, 1952 (Maccabee, ibid.).

105 *"any conceivable threat to the United States"* The press conference was filmed and General Samford's opening statement has been shown in numerous documentaries. It can be seen in the James Fox film *I Know What I Saw* and on this 1952 news clip: http://www.youtube.com/watch?v=utX5HvMO0PM.

105 *"or known types of aerial vehicles"* H. Marshall Chadwell, memorandum for Director of Central Intelligence, December 2, 1952.

105 *"to minimize risk of panic"* H. Marshall Chadwell, memorandum for Director of
Central Intelligence, "Flying Saucers," September 11, 1952, pp. 3–4.

105 *"to review and appraise the available evidence"* "Unidentified Flying Objects,"
December 4, 1952, IAC-M-90.

106 *"the aura of mystery they have unfortunately acquired"* F. C. Durant, "Report of Meetings of Scientific Advisory Panel on Unidentified Flying Objects," convened by
Office of Scientific Intelligence, CA, January 14–18, 1953.

107 *"It made the subject of UFOs scientifically unrespectable"* Hynek, *The Hynek UFO Report*,
p. 23.

108 *"to decide the nature of the UFO phenomenon"* J. Allen Hynek, *The UFO Experience*
(Marlowe & Company, 1998; originally published 1972), p. 169.

108 *"any basis in fact"* Ibid., p. 186.

108 *"as poor as they were"* Ibid., p. 183.

109 *hosted by the trusted Walter Cronkite* This letter, dated September 10, 1966, was
found in the archives of the Smithsonian Institution by Dr. Michael Swords.

109 *"science is more served by fact"* "UFO: Friend, Foe or Fantasy?" hosted by Walter Cronkite, CBS special, 1966, http://www.cbsnews.com/video/watch/?id=
2935380n.

110 *congressional hearings on the subject of UFOs* Congressman Gerald R. Ford, letter to
L. Mendel Rivers, Chairman, Science and Astronautics Committee of the Committee on Armed Services, March 28, 1966; David Michael Jacobs, *The UFO Controversy in America* (Indiana University Press, 1975), p. 204.

111 *"an almost zero expectation of finding a saucer"* Robert J. Low, memo to E. James
Archer and Thurston E. Manning, "Some Thoughts on the UFO Project,"
August 9, 1966, contained in David R. Saunders and R. Roger Harkins, *UFOs?
Yes! Where the Condon Committee Went Wrong* (Signet Books/New American
Library, 1968), pp. 242–44.

111 *"reach a conclusion for another year"* John Fuller, "Flying Saucer Fiasco," *Look*,
May 14, 1968.

112 *what I could call "irrefutable proof"* Hearings before the Committee on Science
and Astronautics, U.S. House of Representatives, Ninetieth Congress, "Symposium on Unidentified Flying Objects," July 29, 1968 (U.S. Government Printing
Office, Washington 1968), p. 32.

112 *cooperation be sought through the United Nations* Ibid., p. 15.

113 *"within the sight of two witnesses"* Edward U. Condon, project director, and Daniel S.
Gillmor, editor, *Scientific Study of Unidentified Flying Objects* (Bantam, 1969), p. 407.

113 *it concluded seven weeks later* "Review of the University of Colorado Report on Unidentified Flying Objects by a Panel of the National Academy of Sciences," 1969.

113 *"got involved in such foolishness"* "Air Force Closes Study of UFO's," *New York Times,* December 18, 1969.

113 *"should arouse sufficient curiosity to continue its study"* J. P. Kuettner et al., "UFO: An Appraisal of the Problem, a Statement by the UFO Subcommittee of the AIAA," *Astronautics and Aeronautics,* 8, no. 11.

CHAPTER 12: TAKING THE PHENOMENON SERIOUSLY

116 *would still be dealt with accordingly* BBC News, "UFO Investigations Unit Closed by Ministry of Defence," December 4, 2009. http://news.bbc.co.uk/2/hi/uk _news/8395473.stm.

116 *now called GEIPAN* GEIPAN stands for Groupe d'Étude et d'Information sur les Phénomènes Aérospatiaux Non-Identifiés (Group for the Study of and Information on Unidentified Aerospace Phenomena).

116 *known as CNES* CNES stands for the Centre National d'Études Spatiales (National Center of Space Studies).

118 *and the Condon report in 1968* Associated Press, "French Space Agency Puts UFO Files Online," March 23, 2007, http://www.foxnews.com/story/0,2933,260590,00 .html.

118 *traditionally employed by that noted paper* Sarah Lyall, "British U.F.O. Shocker! Government Officials Were Telling the Truth," *New York Times,* May 26, 2008.

118 *by former UK Ministry of Defence official Nick Pope* Nick Pope, "Unidentified Flying Threats," *New York Times,* July 29, 2008.

120 *Phénomènes aérospatiaux non identifiés* The book was published by Le Cherche Midi, 2007.

121 *"to identify what we don't know"* http://www.eeb.org/publication/1999/eeb _position_on_the_precautionar.html. See also http://ec.europa.eu/dgs/health _consumer/library/pub/pub07_en.pdf].

CHAPTER 13: THE BIRTH OF COMETA

123 *"to do the research, to work together"* James Fox's film *I Know What I Saw* includes some clips of this interview with General Letty at his home, and it also covers the COMETA Report and the work of GEIPAN.

123 *I first became aware* Oscar Zambrano translated some sections about Captain Girard and Captain Fartek. The rest was written in English.

127 *our responsibility to study them seriously* Interview with General Thouverez, *Armées d'aujourd'hui* (*Armies of Today*), July 2002.

CHAPTER 14: FRANCE AND THE UFO QUESTION

128 *For twenty-one years* The much longer first draft of this piece was written in French and translated by Jean-Luc Rivera. Throughout the editing process, M. Velasco and I worked in English.

129 *the incidents at Malmstrom Air Force Base* Velasco is referring to the 1967 case described by Robert Salas in chapter 15 (pp. 144–45) and other sightings that took place in the Malmstrom area around the same time period.

129 *a new internal agency then called GEPAN* GEPAN: Groupe d'Étude des Phénomènes Aérospatiaux Non-Identifiés (Group for the Study of Unidentified Aerospace Phenomena).

131 *a new agency called SEPRA* SEPRA: Service d'Expertise des Phénomènes de Rentrées Atmosphériques (Service of Expertise on the Phenomena of Atmospheric Reentries).

133 *might account for the chlorophyll reductions* The case was presented in the GEPAN report Note Technique No. 16, Enquete 81/01, "Analyse d'une Trace" (Analysis of Trace Evidence), March 1, 1983.

For more on the Trans-en-Provence case, see "Report on the Analysis of Anomalous Physical Traces: The 1981 Trans-en-Provence UFO Case," by Jean-Jacques Velasco, p. 27, and "Return to Trans-en-Provence," by Jacques F. Vallee, p. 19, in *Journal of Scientific Exploration,* vol. 4, no. 1, 1990. Both articles can also be found in the excellent book *The UFO Enigma: A New Review of the Physical Evidence,* by Peter A. Sturrock (Warner Books, 1999), pp. 257–97.

Vallee's paper is noteworthy: The site of the 1981 Trans-en-Provence UFO case was visited again during 1988. Soil samples taken at the time of the initial investigation were analyzed in an American laboratory in an effort to validate the GEPAN/CNES study of the case. The results of the interviews with the witness and his wife and the examination of samples taken at the surface and below the surface of the physical trace support the findings of the CNES team and the truthfulness of the witness's testimony.

136 *an outstanding independent French investigator* Dominique Weinstein, "Unidentified Aerial Phenomena: Eighty Years of Pilot Sightings—Catalog of Military, Airliner, Private Pilots' Sightings from 1916 to 2000," February 2001, 6th edition.

137 *"there is cause for concern"* Richard Mandelkorn, Commander, U.S. Navy, "Report of Trip to Los Alamos, New Mexico, 16 February 1949," Subject: Project Grudge, February 18, 1949, p. 4. Project Blue Book file.

137 *"It is felt that these incidents"* Memo from Headquarters Fourth Army to Director of Intelligence, "Unconventional Aircraft (Control No. A-1917)," by Colonel Eustis L. Poland. www.project1947.com/gfb/poland.htm.

137 *"the National Defense of the United States"* Report concerning a conference held on April 27 and 28, 1949, at Kirtland Air Force Base on unidentified aerial phenomena, for the director of special investigations, USAF, Washington, D.C., May 12, 1949, p. 4. From Project Blue Book files. Richard Mandelkorn, Commander, U.S. Navy, "Report of Trip to Los Alamos, New Mexico, 16 February 1949," Subject: Project Grudge, 18 February 1949, p. 4. Project Blue Book file.

138 *"and return to home base"* George E. Valley, "Some Considerations Affecting the Interpretation of Reports of Unidentified Flying Objects," report for Project Sign, USAF, originally classified Secret.

139 *and Pease AFB (New Hampshire)* Larry Hatch, Nuclear Connection Project (1998); also see the book by Robert Hastings, *UFOs and Nukes: Extraordinary Encounters at Nuclear Weapons Sites* (AuthorHouse, 2008), for details of these and other incidents.

CHAPTER 15: UFOS AND THE NATIONAL SECURITY PROBLEM

142 *"sightings categorized as 'unidentified' are extraterrestrial vehicles"* The Air Force Fact Sheet, "Unidentified Flying Objects and Air Force Project Blue Book," can be found at http://www.af.mil/information/factsheets/factsheet.asp?fsID =188. The December 17, 1969, News Release, no. 1077–69, "Air Force to Terminate Project Blue Book," was issued by the Office of Assistant Secretary of Defense (Public Affairs), Washington, D.C.–20301. See http://www.dod.gov/pubs/foi/ufo/asdpa1.pdf.

142 *Astronautics Committee in its 1968 hearing* This is the same hearing discussed in chapter 11 in reference to the testimony of James E. McDonald, held just before the Condon report was issued and Project Blue Book was shut down.

142 *"scientific secrets we do not know ourselves"* Hearings before the Committee on Science and Astronautics, U.S. House of Representatives, Ninetieth Congress, "Symposium on Unidentified Flying Objects," July 29, 1968 (U.S. Government Printing Office, Washington 1968), pp. 121–24.

142 *General Samford in his 1952 press conference* See chapter 11 to review details of Samford's press conference. He stated the percentage of credible reports of UFOs did not represent "any conceivable threat to the United States."

143 *"responsibility for the defense of the United States"* UPI, "Air Force Order on 'Saucers' Cited; Pamphlet by the Inspector General Called Objects a 'Serious Business,' " *New York Times*, February 28, 1960.

144 *"UFOs could not be Soviet machines"* Statement of Hon. Leonard G. Wolf of Iowa in the House of Representatives, August 31, 1960. Entered into the Congressional Record, p. 18955.

144 *"cause for grave concern to this headquarters"* SAC headquarters to Ogden Air Material Area (OOAMA) Hill AFB, Utah, "Loss of Stategic Alert, Echo Flight, Malstrom AFB," March 17, 1967. Originally classified Secret. The document is reprinted on p. 108 of Robert Salas and James Klotz, *Faded Giant* (privately published, 2004), a book with useful information about the Malmstrom case and other missile UFO incidents from the 1960s. For a more detailed and broader look at such cases, see Robert Hastings, *UFOs and Nukes: Extraordinary Encounters at Nuclear Weapons Sites* (AuthorHouse, 2008).

145 *directly injected into the equipment* Salas, op. cit., p. 29.

145 *have come forward with similar reports* Hastings, op. cit.

146 *"reports which fall within their responsibilities"* C. H. Bolender, Brig. Gen. USAF, memorandum, "Unidentified Flying Objects (UFO)," October 20, 1969. Obtained through the FOIA by Robert Todd in 1979. http://www.nicap.org/directives/Bolender_Memo.pdf.

147 *the NORAD log reports* 24th NORAD Region Senior Director's Log, November 1975; NORAD Command Director's Log, November 1975.

148 *the* Post *reported* Ward Sinclair and Art Harris, "UFOs Visited U.S. Bases, Reports Say," *Washington Post,* 1979.

150 *This highly unusual report* JCS Communication Center of the USDAQ Tehran Message 230630Z, September 1976, released in 1977 through the Department of Defense, "Reported UFO Sighting," 3 pages plus 1 page evaluation. See also Henry S. Shields, "Now You See It, Now You Don't," United States Air Force Security Service, *MIJI Quarterly* Report 3–78, October 1978.

150 *"maneuverability was displayed by the UFOs"* This list is exactly the way it is in the document, except I've put bullets where there were letters (a.–f.) and removed 1) from the first line ("an outstanding report . . .") to make it easier to read.

151 *Titled "UFO Sighted in Peru"* Department of Defense, Joint Chiefs of Staff message center, "UFO Sighted in Peru," June 3, 1980.

CHAPTER 16: "A POWERFUL DESIRE TO DO NOTHING"

154 *twelve officers from this department alone had their own sightings* Dr. J. Allen Hynek, Philip J. Imbrogno, and Bob Pratt, *Night Siege: The Hudson Valley UFO Sightings* (Llewellyn Publications, 1998), p. 81.

155 *"as if the ball were trying to measure something"* Interview with Heinrich Nicoll for the NBC television series "Unsolved Mysteries," hosted by Robert Stack.

155 *"One came back and the other didn't"* Athens *reported* Hynek, Imbrogno, and Pratt, op. cit., p. 117.

155 *"probing the water"* Ibid., p. 2.

157 *bigger than a football field* Ibid., pp. 165–66.

158 *well-documented visits by some kind of phenomenon* Hynek's research on the Hudson Valley wave was collected into the very readable book *Night Siege*, published after his death, first in 1987 and later reissued, in collaboration with Philip Imbrogno and Bob Pratt.

158 *In a 1985 essay* While he was investigating the Westchester County "boomerangs" in New York, Dr. Hynek left this essay on a diskette at the home of his friend Dr. Willy Smith on August 30, 1985. Titled "The Roots of Complacency," it was meant to be a draft preface for *Night Siege*. A few weeks after writing the piece, Hynek went into surgery. His health rapidly declined in the ensuing months, and he died in April 1986. This last essay is quite different from the much shorter preface to *Night Siege* in its passion and intimate, unedited style.

160 *"springboard to mankind's outlook on the universe"* J. Allen Hynek, *The Hynek UFO Report* (Dell Publishing, 1977), p. 1.

CHAPTER 17: THE REAL X-FILES

163 *"until some material evidence becomes available"* The working party's conclusions, titled "Unidentified Flying Objects" and classified Secret Discreet, were presented in a document dated June 1951, bearing the designation DSI/JTIC Report No. 7. Its six pages are posted at http://www.nickpope.net/documents.htm.

169 *had* not *been released to the public* It was later released in 2001 under the title "Unidentified Flying Objects (U.F.O.'s) Report of Sighting, Rendlesham Forest, December 1980." The key documents are posted at http://www.nickpope.net/documents.htm.

169 *some rapid sketches in his police notebook* A detailed account of what happened is provided in Jim Penniston's contribution to this book, in the next chapter.

172 *one of the most significant UFO sightings ever* For a detailed account of the case, see the book by Georgina Bruni, *You Can't Tell the People* (Pan Books, 2001).

173 *"repeatedly denied, in precisely those terms"* Letter from Lord Hill-Norton to Lord Gilbert, Minister of State, Ministry of Defence, dated October 22, 1997.

175 *"to out-maneuver a UAP during interception"* Defense Intelligence Analysis Staff Study, December 2000, "Unidentified Aerial Phenomena in the UK Air Defense Region," vol. 1, chapt. 5, p. 4. See http://www.mod.uk/NR/rdonlyres/AB43D483-FF03-44F0-85DE-C4233C7C9F10/0/uap_vol1_ch5_pg4.pdf for the relevant extract.

177 *"and dealt with by RAF fighter aircraft"* BBC News, "UFO Investigations Unit Closed by Ministry of Defence," December 4, 2009, http://news.bbc.co.uk/2/hi/uk _news/8395473.stm.

Chapter 19: Chile

189 *known as the OIFAA* OIFAA stands for Oficina de Investigación de Fenomenos Aéreos Anómalos (Office for the Investigation of Anomalous Activity). The Peruvian Air Force established the agency within the DINAE, Division de Intereses Aeroespaciales (Division of Aerospace Interests), an Air Force department, in December 2001.

189 *physically real but could not be explained* Dr. Anthony Choy, a UAP field investigator and founding member of the OIFAA, was the lead investigator in the very remote Chulucanas region, beginning even before the OIFAA was founded in 2001. Choy had the unique of experience in 2003 of actually witnessing a dramatic UFO event over the town square of an ancient village, along with about forty other witnesses, while in the process of conducting investigations. His studies in this region precipitated the Air Force's first acknowledgment of a physically real but unknown phenomenon. Choy is now petitioning the Peruvian government to declassify its UFO files.

190 *who presides over an Air Force commission studying the cases* Daniel Iglesias, "Uruguay: Air Force Declassifies UFO Files, ET Hypothesis Not Dismissed," June 6, 2009, http://www.elpais.com.uy.

190 *The CEFAA* CEFAA stands for el Comité de Estudios de Fenómenos Aéreos Anómalos (the Committee for the Study of Anomalous Aerial Phenomena). It was established in October 1997 within the Department of Civil Aeronautics, the Dirección General de Aeronáutica Civil or DGAC, the Chilean equivalent of the American FAA.

190 *Bermúdez wrote in an e-mail* These details about the ceremony come from a personal e-mail communication with General Bermúdez in January 2010.

190 *In the last days of March* Sections of this piece were translated from the Spanish by Gustavo Rodríguez Navarro, Oscar Zambrano, and Andrea Soares Berrios.

194 *The aviation branch of the Army of Chile* Brigada de Aviación del Ejército de Chile, the aviation branch of the Army of Chile, is known as BAVE.

Chapter 20: UFOs in Brazil

199 *by the Navy, which compiled a report* Carlos Alberto Ferreira Bacellar, Commander of the Oceanographic Station at Trindade, "Clarification of the Observation of Unidentified Flying Objects Sighted on the Island of Trindade, in the Period of 12/5/57 to 1/16/58."

199 *experts in America conducted further analysis* The information on the Trindade pho-
     tos was obtained from Don Berliner, *UFO Briefing Document* (Dell Publishing,
     1995), pp. 71–77. This report is available at http://www.bibliotecapleyades.net/
     ciencia/ufo_briefingdocument/1958.htm#50.

199 *many of them concerning the Air Force's "Operation Saucer"* The Brazilian files
     can be viewed online. See the National Intelligence Agency report of Opera-
     tion Saucer: http://www.ufo.com.br/public/prato/ACE_3370.83.pdf; also, the
     Brazilian Air Force report of Operation Saucer: http://www.ufo.com.br/public/
     prato/ACE_3370.83.pdf. More recent files can be accessed at www.ufo.com
     .br/public/abertura_2; and more through these links: www.ufo.com.br/public/
     abertura_1, www.ufo.com.br/public/brasil, www.ufo.com.br/public/documents,
     and www.ufo.com.br/public/prato.

199 *All was translated from Portuguese* Translation services were provided by Eduardo
     Rado of Brazil and Andrea Soares Berrios of New York.

200 *"to fly in formation, not necessarily manned"* Air Brigadier José Pessoa Cavalcanti de
     Albuquerque, commander of the Aerial Command of Aerial Defense, to Aerial
     General Command, "Occurrence Report," June 2, 1986. Both the original docu-
     ment in Portuguese and an English translation can be seen at www.ufo.com
     .br/documentos/night.

200 *Brazilian Airspace Defense Command* Comando de Defesa Aérea Brasileiro, known
     as COMDABRA.

202 *to the air defense operations center* Comando de Operações de Defesa Aérea, known
     as CODA.

CHAPTER 21: FIGHTING BACK

212 *"Unidentified Flying Objects and related phenomena"* Dept. of State Teletype, "Gre-
     nadian UFO Crusade: Déjà vu," November 18, 1978, released through the Free-
     dom of Information Act. Sourced from Clifford E. Stone, *UFOs Are Real* (SPI
     Books, 1997).

213 *"and scientists, including astronomers"* Dept. of State Teletype, "Grenadian UFO
     Resolution," November 28, 1978, classified Confidential and released through
     the FOIA. Sourced from Clifford E. Stone, *UFOs Are Real* (SPI Books, 1997),
     Doc. 5–21a.

213 *a "blitzkrieg sales pitch"* Dept. of State Teletype, November 18, 1978, op. cit.

213 *"U.S. nationals on the Grenadian delegation"* Ibid.

213 *"and gamble on the results"* Dept. of State Teletype, "Grenadian UFO Resolution,"
     December 2, 1978. Sourced from Clifford E. Stone, *UFOs Are Real* (SPI Books,
     1997), Doc. 5–22.

213 *"of this French investigation are profound"* J. Allen Hynek, speech to the United

Nations, November 27, 1978. His speech is summarized in a State Dept. Teletype, "Grenadian UFO Resolution," November 28, 1978.

214 *then it "disappeared"* Sighting Report for the International UFO Bureau, Oklahoma City, Oklahoma, September 18, 1973.

214 *"ought to be NASA"* Dr. Frank Press, letter to Dr. Robert Frosch, July 21, 1977. Richard C. Henry, "UFOs and NASA," *Journal of Scientific Exploration,* vol. 2, no. 2 (1988), p. 109.

214 *NASA name a "project officer"* Dr. Robert Frosch, letter to Dr. Frank Press, September 6, 1977. Full letter included in the appendix: Richard C. Henry, "UFOs and NASA," *Journal of Scientific Exploration,* vol. 2, no. 2 (1988), pp. 110–11.

214 *White House concurred without delay* Dr. Frank Press, letter to Dr. Robert Frosch, September 14, 1977. Full letter included in the appendix: Richard C. Henry, "UFOs and NASA," *Journal of Scientific Exploration,* vol. 2, no. 2 (1988), p. 114.

214 *"preventing a reopening of UFO investigations"* Charles E. Senn, letter to Duward L. Crow, September 1, 1977.

215 *"to convene a symposium on this subject"* Dr. Robert Frosch, letter to Dr. Frank Press, December 21, 1977. Henry, op. cit., p. 115.

215 *"to intelligences far beyond our own"* Dr. Richard Henry, memorandum to Dr. Noel Hinners, Subject: UFO Matters, January 17, 1978; Richard C. Henry, "UFO's and NASA," *Journal of Scientific Exploration,* vol. 2, no. 2 (1988), p. 130.

216 *first step in conducting any sort of investigation* See chapter 6 about the O'Hare incident in 2006 and the FAA response.

216 *which was reported by the BBC* See chapter 7 by way of review.

216 *"confounded local military experts and local police"* BBC News, "UFO Baffles Aviation Experts," September 15, 1999, http://news.bbc.co.uk/2/hi/uk_news/ 448267.stm.

217 *The 2010 FAA Aeronautical Information Manual* See section 7–6-4, "Unidentified Flying Object (UFO) Reports." The manual can be viewed at http://www.faa .gov/air_traffic/publications/ATpubs/AIM/Chap7/aim0706.html.

217 *"report the activity"* FAA manual, ibid.

218 *Ministry of Defence monitors them for that reason* Author interview with Nick Pope, via a series of e-mails, August 2009.

220 *a spokesman explained at the time* Associated Press, "FAA Investigates JAL Flight 1628 UFO Sighting," 1986; and United Press International, "Pilot Describes 'Unbelievable' UFO Encounter," December 31, 1986.

220 *As revealed in a verbatim transcript* "Record of Interview with JAL Captain," January 2, 1987, pp. 16–17. The transcript was provided to researchers by the FAA in 1987.

221 *"was unable to confirm the event"* Bruce Maccabee, "The Fantastic Flight of JAL 1628," http://brumac.8k.com/JAL1628/JL1628.html. This is the most complete report about the sighting over Alaska, and highly recommended. Dr. Maccabee is the author or co-author of about three dozen technical articles and more than a hundred UFO articles over the last twenty-five years. He is also a leading photo-analyst of UFO images. See http://brumac.8k.com.

221 *during the FAA's interviews with these witnesses* AP and UPI, op. cit.

CHAPTER 22: THE FAA INVESTIGATES A UFO EVENT "THAT NEVER HAPPENED"

227 *"that we were visited by a UFO"* A Twix is a message sent to all media advising them to broadcast or print a news article. It can be transmitted either by e-mail or fax or as a print-out.

CHAPTER 23: GOVERNMENT COVER-UP

238 *even "no comment" is considered a confirmation* Sweetman is the North American editor for *Jane's Defence Weekly*. This story was published in *Jane's International Defence Review* in 2000; see http://www.janes.com/defence/news/jidr/jidr000105_01_n.shtm.

243 *"classified above Top Secret," he wrote in a 1975 letter* Senator Barry Goldwater, letter to "Mr. S A," on "United States Senate" letterhead, March 28, 1975. The name of the recipient, who had written to inquire about the senator's interest in UFOs, was redacted from the document when it was released through the Freedom of Information Act. The names on numerous other similar letters were not redacted.

243 *"it is just impossible to get anything on it"* Senator Barry Goldwater, letter to Lee M. Graham on "United States Senate" letterhead, October 19, 1981.

243 *"I've never tried to make it my business since"* Senator Barry Goldwater, letter to William S. Steinman, on "United States Senate" letterhead, June 20, 1983.

243 *" 'Don't ever ask me that question again!' "* The radio clip can be viewed on YouTube at http://www.youtube.com/watch?v=gPFBg1NNUBU. Numerous websites state that this clip was an excerpt from a 1994 interview with Larry King on CNN, but I have not been able to verify this.

244 *"who destroyed the messages, or why"* News Release, U.S. Congressman Steve Schiff, First Congressional District New Mexico, July 28, 1995.

244 *possible Soviet nuclear testing* James McAndrew, Headquarters United States Air Force, "The Roswell Report: Case Closed," July 1994, http://www.af.mil/information/roswell/index.asp.

CHAPTER 24: GOVERNOR FIFE SYMINGTON AND MOVEMENT
TOWARD CHANGE

248 *watching it take off in the blink of an eye* NUFORC reports can be found at
www.nurforc.org. See also the feature-length documentary *I Know What I Saw,*
directed by James Fox, for witness interviews (trailer can be seen at www.iknow
whatisawthemovie.com).

249 *into the national spotlight* Richard Price, "Arizonans Say the Truth about UFOs Is
Out There," *USA Today,* June 18, 1997.

249 *"find out if it was a UFO"* The television clip from a local Arizona station
is included in *I Know What I Saw.*

249 *mask was removed before the cameras* A clip from television coverage of the press
conference is featured in *I Know What I Saw.*

250 *which could be seen for 150 miles* Senator John McCain, letter to constituent (name
redacted), United States Senate, October 9, 1997.

250 *were, in fact, flares* Dr. Bruce Maccabee, "Report on Phoenix Light Arrays," 1998,
http://brumac.8k.com/phoenixlights1.html. This detailed study concludes with
the following:

> "The most parsominous explanation for these lights is that they were flares
> (as so stated for the March 13, 1997, lights by the Maryland National Guard).
> This analysis is therefore consistent with that of the Cognitech Corporation
> (Dr. Leonid Rudin) done for the Discovery Channel documentary (November,
> 1997). It is also consistent with the analysis of Dr. Paul Scowen, professor of
> astronomy at ASU, as reported by author Tony Ortega in the Phoenix "New
> Times" newspaper, March 5–11, 1998, which showed that the lights were far-
> ther away than the mountain peaks in the K video. In that newspaper article
> the author also reported that an "Arizona National Guard public information
> officer, Captain Eileen Benz, had determined that the flares had been dropped
> at 10 P.M. over the North Tac Range 30 miles southwest of Phoenix at an unusu-
> ally high altitude of 15,000 ft." Except for the stated distance, which should be
> more like 60 miles (and up to 100 miles away), this statement is consistent with
> the analysis presented here."

> A second paper by Dr. Maccabee, "Supplementary Discussions of the Phoe-
> nix Lights Videos of March 13, 1997," January, 2006, can be found at http://
> brumac.8k.com/PhoenixSupplement/.

250 *"any evidence whatsoever of aliens or UFOs"* Dennis Roberts, reporter for the *Modesto
Bee,* a Northern California newspaper, attended the press conference in Stockton,
California, and taped it. He sent me a transcript in an e-mail on March 1, 2000.

251 *over the Royal Air Force base at Cosford* See chapter 17 by Nick Pope, "The Real
X- Files," for a review of the Cosford incident.

260 *as Goldwater has written in his letters* See the previous chapter for excerpts from these letters.

CHAPTER 26: ENGAGING THE U.S. GOVERNMENT

265 *I cofounded the Coalition for Freedom of Information* See www.freedomofinfo.org for more information on the Coalition for Freedom of Information, CFi.

266 *as a way of saving face* MSNBC presidential debate, October 30, 2007. Transcript: http://www.msnbc.msn.com/id/21528787/page/22/.

268 *"for for the benefit of science"* Go to www.freedomofinfo.org for the full text of the International Declaration to the United States Government, first released in November 2007.

CHAPTER 27: MILITANT AGNOSTICISM AND THE UFO TABOO

269 *the overwhelming evidence for its existence* Alexander Wendt and Robert Duvall, "Sovereignty and the UFO," *Political Theory,* vol. 36, no. 4 (August 2008), pp. 607–33. Sage Publications has posted the paper on its website, http:ptx.sagepub.com.

269 *a bad fairy had administered a sleeping potion* Hynek, "The Roots of Compacency," 1985, op. cit.

272 *UFO skeptics think that human beings know* The widely used phrase "UFO skeptic" can be misleading, because "skepticism" should imply doubt but openness. However, in UFO discourse it has been deformed into positive denial.

273 *over 400 extrasolar planets* Dennis Overbye, "A Sultry World Is Found Orbiting a Distant Star," *New York Times,* December 17, 2009.

273 *"They Can't Get Here"* Some of this section is reprinted verbatim from the 2008 paper, Alexander Wendt and Raymond Duvall, op. cit., p. 616. Occasional phrases or sentences in "Militant Agnosticism and the UFO Taboo" were also used in the first paper.

273 *civilizataions should have reached Earth long ago* Martyn Fogg, "Temporal Aspects of the Interaction Among the First Galactic Civilizations," *Icarus* 69 (1987): 370–84.

273 *the speed of light is truly an absolute barrier* J. Deardorff et al., "Inflation-Theory Implications for Extraterrestrial Visitation," *Journal of the British Interplanetary Society* 58 (2005): 43–50.

274 *to skip over space without time dilation* H. E. Puthoff, S. R. Little, and M. Ibison, "Engineering the Zero-Point Field and Polarizable Vacuum for Interstellar Flight," *Journal of the British Interplanetary Society* 55 (2002): 137–44.

274 *"Fermi Paradox"* Stephen Webb, *Where Is Everybody?* (New York: Copernicus Books, 2002).

277 *UFOs are not a national security concern* Richard Dolan, *UFOs and the National Security State,* pp. 193–203.

278 *reinforcing the skeptical case* Peter Galison, "Removing Knowledge," *Critical Inquiry* 31 (2004): 229–43. On UFO secrecy see especially Dolan, *UFOs and the National Security State,* and, for the official view, Gerald Haines, "CIA's Role in the Study of UFOs, 1947–1990," *Intelligence and National Security* 14 (1999): 26–49, and Charles Ziegler, "UFOs and the US Intelligence Community," *Intelligence and National Security,* vol. 14 (1999), pp. 1–25.

281 *a science of UFOs* One could imagine, for example, a complementary, bottom-up or "democratic" strategy centering on an Internet-funded NGO, an idea we (Wendt and Duvall) have explored elsewhere.

CHAPTER 28: FACING AN EXTREME CHALLENGE

284 *"a powerful desire to do nothing"* Hynek, "The Roots of Complacency," op. cit.

284 *"the dam breaks, sometimes cataclysmically"* Ibid.

286 *Dr. James E. McDonald, senior atmospheric physicist* Dr. James E. McDonald, "Statement on Unidentified Flying Objects," submitted to the House Committee on Science and Astronautics, Symposium on Unidentified Flying Objects, Washington, D.C., July 29, 1968. This report is recommended reading. For a detailed biography about McDonald and chronicle of his work, see Ann Druffel, *Firestorm: Dr. James E. McDonald's Fight for UFO Science* (Wild Flower Press, 2003).

286 *a 2009 memoir* Peter Sturrock, *A Tale of Two Sciences: Memoirs of a Dissident Scientist* (Exoscience, 2009).

287 *on the position of the scientific establishment* Peter Sturrock, *The UFO Enigma: A New Review of the Physical Evidence* (Warner Books, 1999). Provides all the case reports presented at the conference. Recommended reading.

287 *"If it does happen, it can happen."* Ibid., p. 160.

290 *multiple dimensions, and even time travelers* See Michio Kaku, *Physics of the Impossible: A Scientific Exploration into the World of Phasers, Force Fields, Teleportation, and Time Travel* (Doubleday, 2008).

290 *an August 2009* Newsweek *cover story* Andrew Romano, "Aliens Exist," *Newsweek,* August 24 & 31, 2009, pp. 50–52.

290 *over 400 planets orbiting other stars* Marc Kaufman, "Search for Extraterrestrial Life Gains Momentum Around the World," *Washington Post,* December 22, 2009. In addition to the exoplanets already discovered, the article states, "It is generally assumed that billions or trillions more are orbiting in distant systems."

290 *NASA has also developed a highly sensitive infrared space telescope* NASA release, "NASA's Wise Eye Spies First Glimpse of the Starry Sky; Infrared All-Sky Surveying Telescope Sends Back First Images from Space," January 6, 2010. It begins, "NASA's Wide-field Infrared Survey Explorer, or WISE, has captured its first look at the starry sky that it will soon begin surveying in infrared light. Launched

on Dec. 14, WISE will scan the entire sky for millions of hidden objects, including asteroids, 'failed' stars and powerful galaxies." More information about the WISE mission is available at: http://www.nasa.gov/wise.

292 *"An extraordinary phenomenon demands an extraordinary investigation."* UFO researcher and author Budd Hopkins originated this phrase in 1987 while conversing with astronomer Carl Sagan in the greenroom of a Boston TV station. For an account of the interchange, see Budd Hopkins, *Art, Life and UFOs: A Memoir* (Anomalist Books, 2009).

INDEX